Leila Esmaeili Sereshki

Modeling diffusion controlled reactions in living cells

Leila Esmaeili Sereshki

Modeling diffusion controlled reactions in living cells

A solution to the subdiffusion-efficiency paradox of EcoRV enzyme in E.coli bacteria

Südwestdeutscher Verlag für Hochschulschriften

Impressum/Imprint (nur für Deutschland/only for Germany)
Bibliografische Information der Deutschen Nationalbibliothek: Die Deutsche Nationalbibliothek verzeichnet diese Publikation in der Deutschen Nationalbibliografie; detaillierte bibliografische Daten sind im Internet über http://dnb.d-nb.de abrufbar.
Alle in diesem Buch genannten Marken und Produktnamen unterliegen warenzeichen-, marken- oder patentrechtlichem Schutz bzw. sind Warenzeichen oder eingetragene Warenzeichen der jeweiligen Inhaber. Die Wiedergabe von Marken, Produktnamen, Gebrauchsnamen, Handelsnamen, Warenbezeichnungen u.s.w. in diesem Werk berechtigt auch ohne besondere Kennzeichnung nicht zu der Annahme, dass solche Namen im Sinne der Warenzeichen- und Markenschutzgesetzgebung als frei zu betrachten wären und daher von jedermann benutzt werden dürften.

Coverbild: www.ingimage.com

Verlag: Südwestdeutscher Verlag für Hochschulschriften GmbH & Co. KG
Heinrich-Böcking-Str. 6-8, 66121 Saarbrücken, Deutschland
Telefon +49 681 37 20 271-1, Telefax +49 681 37 20 271-0
Email: info@svh-verlag.de

Approved by: München, TU, Diss., 2012

Herstellung in Deutschland (siehe letzte Seite)
ISBN: 978-3-8381-3398-0

Imprint (only for USA, GB)
Bibliographic information published by the Deutsche Nationalbibliothek: The Deutsche Nationalbibliothek lists this publication in the Deutsche Nationalbibliografie; detailed bibliographic data are available in the Internet at http://dnb.d-nb.de.
Any brand names and product names mentioned in this book are subject to trademark, brand or patent protection and are trademarks or registered trademarks of their respective holders. The use of brand names, product names, common names, trade names, product descriptions etc. even without a particular marking in this works is in no way to be construed to mean that such names may be regarded as unrestricted in respect of trademark and brand protection legislation and could thus be used by anyone.

Cover image: www.ingimage.com

Publisher: Südwestdeutscher Verlag für Hochschulschriften GmbH & Co. KG
Heinrich-Böcking-Str. 6-8, 66121 Saarbrücken, Germany
Phone +49 681 37 20 271-1, Fax +49 681 37 20 271-0
Email: info@svh-verlag.de

Printed in the U.S.A.
Printed in the U.K. by (see last page)
ISBN: 978-3-8381-3398-0

Copyright © 2012 by the author and Südwestdeutscher Verlag für Hochschulschriften GmbH & Co. KG and licensors
All rights reserved. Saarbrücken 2012

TECHNISCHE UNIVERSITÄT MÜNCHEN
Physik-Department T30 g

Modeling diffusion controlled reactions in living cells

Leila Esmaeili Sereshki

Vollständiger Abdruck der von der Fakultät für Physik
der Technischen Universität München
zur Erlangung Grades eines
Doktors der Naturwissenschaften (Dr. rer. nat.)
genehmigten Dissertation.

Vorsitzender: Univ.-Prof. Dr. Friedrich C. Simmel

Prüfer der Dissertation:

         1. Univ.-Prof. Dr. Ralf Metzler, Universität Potsdam
         2. Univ.-Prof. Dr. Martin Zacharias

Die Dissertation wurde am 09.02.2012 bei der Technischen Universität München eingereicht und durch die Fakultät für Physik am 05.03.2012 angenommen.

Abstract

Macromolecular crowding in living biological cells effects subdiffusion of larger biomolecules such as proteins and enzymes. Mimicking this subdiffusion in terms of random walks on a critical percolation cluster, fractional Brownian motion and continuous time random walk, we here present a case study of EcoRV restriction enzymes involved in vital cellular defence. EcoRV has been found in two configurational states. The unbound protein switches between an inactive structure with a closed cleft and another, in which the cleft is more open. It is able to cleave and deactivate the foreign DNA, only when it is in its open state. Surprisingly, the probability x_{act} to find the enzyme in the active state at a given instant of time is as low as ∼1%.

We show that due to its so far elusive propensity to an inactive state the enzyme avoids non-specific binding and remains well-distributed in the bulk cytoplasm of the cell. Despite the reduced volume exploration of subdiffusion processes, the low activity of the enzyme surprisingly guarantees a high efficiency of the enzyme.

Analysing different stochastic processes for subdiffusion and by variation of the non-specific binding constant and anomaly we demonstrate that reduced non-specific binding are beneficial for efficient subdiffusive enzyme activity even in relatively small bacteria cells.

Our results corroborate a more local picture of cellular regulation and demonstrated a solution to subdiffusion-efficiency paradox; Specific molecular design renders the efficiency of EcoRV enzymes almost independent on the exact diffusion conditions. This case study also provides us a chance to compare different subdiffusive processes in living cells and guides us to a broader understating of diffusion controlled reactions in living cells.

Contents

Introduction   5

1 Fractals   11
    1.1 Properties of fractals 13
    1.2 Random fractals 14
    1.3 Percolation 17
       1.3.1 Percolation transition 17
       1.3.2 Structural properties of the percolation 21

2 Diffusion   26
    2.1 Random walks 26
    2.2 The probability density 28
    2.3 Anomalous diffusion 31
       2.3.1 Continuous time random walk 31
       2.3.2 Random walks and diffusion on fractal objects 32
       2.3.3 Diffusion in percolation clusters 33
       2.3.4 Spectral dimension and fractons 34
       2.3.5 Fractional Brownian motion 35
    2.4 First passage time 36
    2.5 Ergodicity 38
       2.5.1 Ergodicity in different processes 39

3 Diffusion in percolation cluster   42
    3.1 Time dependent first passage properties between two spheres   42
    3.2 Time dependent FPT on fractals 46

3.3		Simulation of Random walk	49
	3.3.1	Analysis of the mean squared displacement	50
	3.3.2	Analyzing the first passage time density	54

4 Modeling EcoRV's dynamic in E.coli cell   59

4.1		Theory	62
4.2		Percolation cluster application	64
	4.2.1	Error bars	67
	4.2.2	EcoRV's dynamic's modeled by percolation cluster	68
4.3		FBm application	76
	4.3.1	Generating fractional Brownian motion	77
	4.3.2	Mean squared displacement for fBm	79
	4.3.3	EcoRV's dynamic's modeled by fBm	79
4.4		CTRW application	81
	4.4.1	Generating waiting times	83
	4.4.2	Waiting time distribution with a cutoff	84
	4.4.3	EcoRV's dynamic's modeled by CTRW	85
	4.4.4	CTRW on percolation cluster	87

Conclusions   89

A Generating percolation cluster   92

B Hosking   97

C Waiting time distribution with cutoff   99

Introduction

In many processes such as gelation, coagulation, crystallization, or self-assembly in colloidal or polymer systems, thin film growth in materials science and chemical reactions in biology, the components as the first step perform a random movement in a fluid, diffuse [1] and when they are close together, a reaction may start. The usual paradigm for biochemical reactions assumes the formation of an encounter complex, that may undergo chemical transformation and forms a product. In diffusion limited processes, the fixation step (product formation) proceeds much faster than the diffusion of reactants, and thus the rate is governed by diffusion. Diffusion control is more likely in solution where diffusion of reactants is slower due to the greater number of collisions with solvent molecules. Reactions where the encounter forms easily and the products form rapidly are most likely to be limited by diffusion control. Diffusion limited processes are commonly found in biochemical processes such as catalysis and enzymatic reactions, regulation processes, protein aggregation, and complexation in cells [2].

The simplest model of diffusion-limited encounter in three dimensions has been formulated by Smoluchowski. In Smoluchowski theory reactants are assumed to be noninteracting, spherical, and chemically isotropic. Smoluchowski absorption rate of a diffusing particle of radius R_p and a fixed sphere target of radius R_t (where the annihilation occurs) [3], is

$$\kappa_t = 4\pi D R p_\infty, \qquad (1)$$

where p_∞ is the relative bulk density of the reactant, D is the diffusion coefficient and $R = R_t + R_p$ is the encounter distance.

The Smoluchowski theory is still the main theoretical framework within which the previously mentioned processes are analyzed and this approach is strictly valid only for diluted solutions. Diffusion limited biochemical cellular reactions underlying regulation processes have traditionally been investigated at dilute solvent conditions [4]. Most systems, for instance cells, contain a large number of proteins, nucleic acids, and other smaller molecules that occupy up to 30% to 40% of the available volume [5, 6]. Since no single macromolecular species need to be present at high concentration, such media are referred to as crowded or volume occupied, rather than concentrated [7]. In addition, the macromolecules cannot interpenetrate, therefore the fraction of volume into which a macromolecule can be placed, is much less than the fraction of volume into which a small molecule can be placed (Fig. 1). The total free energy

Introduction

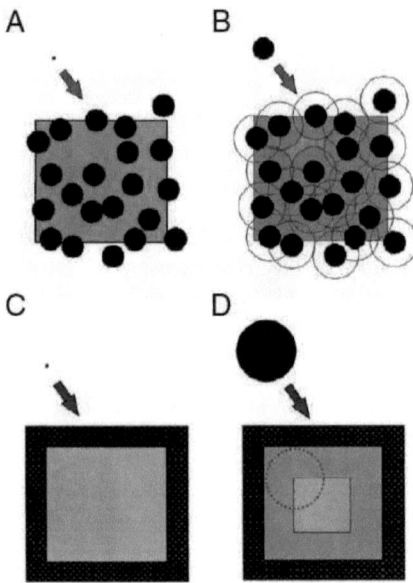

Figure 1: Schematic illustration of available volume (blue) and excluded volume (pink and black) to the center of a small(A,C) and large spherical molecule (B,D) added to a solution containing an approximately 30% volume fraction of large spherical molecules (A,B)[9]

of interaction between a recently added molecule and all the other molecules in the crowded medium is inversely proportional to the probability of placement of that molecule at a random location within the crowded medium [8], and the extra work required to transfer the molecule to a crowded solution resulting from steric repulsion between that molecule and background molecules depends upon its size and shape relative to the background molecules [9, 10] (Fig. 1). It raises the question whether such effects are important for the cell. Zimmerman and Trach estimated that excluded volume effects in the cytoplasm of Escherichia coli (E.coli) bacterium, without well defined nucleus, are comparable to those obtained in a 35% solution of a $\sim$ 70 kDa globular protein, such as bovine serum albumin or hemoglobin [11]. Any cell is extremely large relative to any particular macromolecule and is likely to contain several micro-environments, within each of which a particular macromolecule will be subject to a different set of background interactions. The cytoplasm of E.coli contains at least three such micro-environments: the immediate vicinity of the inner plasma membrane, within which the macromolecule of interest will encounter a high local concentration of membrane phospholipids and proteins, the interior and immediate vicinity of the nucleoid, where the macromolecule will encounter an extremely high local concentration of DNA, and the remaining cytoplasm, within which the macromolecule will be subject mainly to the influence of other soluble proteins, RNAs, and possibly other large particles such as lipid granules.

As a matter of fact, crowding effects are expected to impact profoundly on the thermodynamics and kinetics of biological processes in vivo [5, 12], such as protein folding and stability [13], aggregation [14] and changes in enzyme function and turnover [15].

Larger biopolymers and tracers in living biological cells and artificially crowded control environments in accord with recent high-detail simulations [15] perform subdiffusion of the form [16, 17, 18, 19, 20, 21]

$$\langle \mathbf{r}^2(t) \rangle \simeq t^\alpha \text{ with } 0 < \alpha < 1, \tag{2}$$

as observed experimentally for particles as small as 10 kD, with α in the range of 0.40 to 0.90 [16, 17, 20]. The observed subdiffusion has been measured to persist over tens to hundreds of seconds [16, 17] and thus appears relevant to cellular processes such as gene regulation or molecular defence mechanisms. Subdiffusion leads to reduced global volume exploration and dynamic localisation at reactive interfaces [22]. It is argued that molecular processes in the cell could not be subdiffusive, as this would hinder enzymes activity due to the slowness of the response to external and internal perturbations, and finally compromise the overall fitness of the cell but computer simulations highlighted that the probability to reach a target is increased for a subdiffusive particle as compared to a normal diffusive particle and cell may benefit from the subdiffusion of macromolecules in its interior [16, 23]. In that sense subdiffusion would give rise to a more local picture of diffusion-limited biochemical reactions in biological cells.

In this thesis we present a further clue to understanding the relation between crowding-induced anomalous diffusion and the design of vital cellular mechanisms. Our case study addresses the dynamics of the type II restriction endonuclease EcoRV, that occurs in the bacterium E.coli.

E.coli is a common gram-negative bacterium found in normal human bacterial flora. In fact, the presence of E.coli and other kinds of bacteria within our intestines is necessary to help the human body develop properly and to remain healthy. Some strains, however, can cause severe and life-threatening diarrhea. E.coli cells are typically rod-shaped, and are about 2.0 micrometers (μm) long and 0.5 μm in diameter, with a cell volume of $0.6 - 0.7$ (μm^3) [24] and E.coli DNA is 1.5 mm long, mainly concentrated in the middle of the cell and occupies around a quarter of the cell volume.

One of the enzymes in Escherichia coli is EcoRV (read Eco-R-five), a type II restriction endonuclease. Its name indicates that the restriction endonuclease is found in Escherichia coli ("Eco"), strain RY13 ("R"), restriction endonuclease number "V".

Introduction

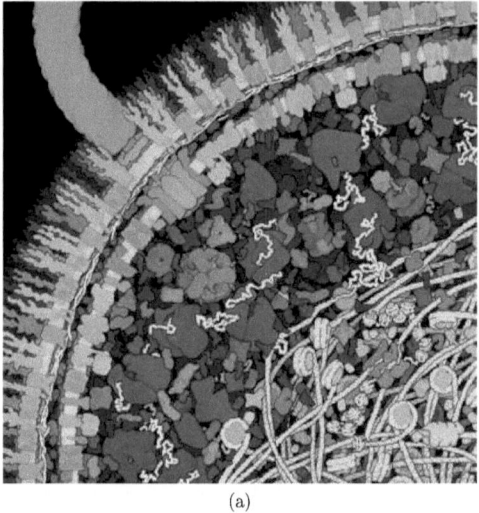
(a)

Figure 2: (a) Artistic view of a slice through an E.coli cell, courtesy David Goodsell, Sripps. In the cytoplasm biomacromolecules such as ribosomes, larger proteins, and messenger RNA are seen.

Restriction enzymes (restriction endonucleases) are found in bacteria and evolved to provide a defense mechanism against invading viruses. They cleave the phosphodiester bond (covalent bonds between a phosphate group and two 5-carbon ring carbohydrates) within a polynucleotide chain (biopolymers like deoxyribonucleic acid (DNA) and ribonucleic acid (RNA) composed of 13 or more nucleotide monomers covalently bonded in a chain). Typically, a restriction site will be four to six nucleotides long. Restriction endonucleases are divided into three categories, Type I, Type II, and Type III, according to their mechanism of action.

EcoRV recognizes the 6-base DNA sequence 5'-GAT|ATC-3' and makes a cut at the vertical line therefore renders it inactive with respect to transcription and replication. EcoRV forms a homodimer in solution before binding and acting on its recognition sequence [25]. Initially the enzyme binds weakly to a non-specific site on the DNA (non-specific binding) and randomly walks along the molecule until the specific recognition site is found [26]. Then it binds to the specific site (specific binding) and cleavage occurs within the recognition sequence and does not require ATP hydrolysis [26]. DNA cleavage is an important mechanism in the cellular defence against foreign DNA of viruses attacking the cell. The cell's own DNA is protected against EcoRV action by methylation by a modification enzyme of the DNA at cytosine or adenine [26]. Bacteria use methylase to be able to differentiate between foreign genetic material and their own, therefore protecting their DNA from their own immune system.

Interestingly, as seen by X-ray crystallography, EcoRV can be found in two configurational states [27, 28]. The unbound protein may switch between an inactive structure with a closed cleft and another, in which the cleft is more open. In open state EcoRV non-specifically binds to DNA that could be the native DNA or foreign DNA.

Remarkably, the probability x_{act} to find the enzyme in the open-cleft, active state at a given instant of time is as low as $\sim 1\%$ [28, 29]. It is a priori puzzling why a vital defence mechanism should be equipped with such a low activity. A physiologic rationale of the open/closed isomerisation could be to reduce non-specific binding to the cell's native DNA. Alien DNA invading the cell would thus immediately be surrounded by a higher EcoRV concentration that, after switching to the active state, could attack this DNA [29]. Here we study the stochastic dynamics of EcoRV in E.coli. We consider possible methods to simulate crowding induced subdiffusion in the cell. Typically there are three prominent physical models for subdiffusion of particles in cells and each corresponds to a distinct potential cellular mechanism:

First, the cytoskeleton is made up of semi flexible polymeric filaments such as microtubules and F-actins, which can be branched and cross-linked by proteins. This scaffold is now considered as a set of fixed obstacles which diffusing particles must navigate. Also, the cytoplasm can be compartmentalized by lipid membranes which further constrain the particle. Such environment with obstacles can be modeled by a random walk on a fractal. The random walker meets the dead ends on all scales and the motion is subdiffusive. The anomalous diffusion exponent is related to the fractal and spectral dimensions, d_f and d_s, characteristics of the fractal, through $\alpha = d_s/d_f$ [30]. A typical example is the subdiffusion on a percolation cluster near criticality that was actually verified experimentally [31]. Recent studies [32] show that the crowded cytoplasm may have a random fractal structure, and bearing in mind that, volume fraction f of the cellular cytoplasm is close to the site percolation on cubic lattice (threshold is $f \approx 31\%$) and that of bond percolation $f \approx 25\%$ [30, 33], molecular crowding may indeed appear severe. To model the cytoplasmic crowding we use static bond percolation cluster, where the bonds connecting the sites of a regular lattice of the d-dimensional space are present with probability p. The ensemble of points connected by bonds is called a cluster. If p is above the percolation threshold p_c, an infinite cluster exists. If $p = p_c$, this infinite cluster is a random fractal characterized by its fractal dimension d_f. Number of recent works applied the percolation idea to stochastic motion in a crowded environment [34, 35, 36, 37].

Second, macromolecular crowding and the presence of elastic elements, such as cytoskeletal filaments and nucleic acids, give the cytoplasm viscoelastic properties. As a particle moves through this medium, the cytoplasm "pushes back", creating long time correlations. This memory leads to subdiffusive behavior that can be modeled by fractional Brownian motion (fBm). In normal Brownian motion on a surface the fractal dimension of the random walk is $d_f = 2$, but in fBm process particle explores more than just a surface as its fractal dimension is given by $d_f = 2/\alpha$ which $\alpha = 2H$ [38]. Thus the sampled subspace is considerably larger than a surface with dimension of $d_f = 2$ and may even exceed the dimension of the bulk ($d_f = 3$).

Third model is the continuous time random walk (CTRW) [39]. If a particle diffusing through the cytoplasm encounters a binding partner, then it will pause for a period of time before dissociating and diffusing away. Multiple binding events with a range of rate constants generate long tails in the waiting time distribution, leading to subdiffusive behavior [39]. Another scenario could be a tracer in a very crowded environment trapped in dynamic "cages" whose life times are broadly distributed. In particular, the CTRW induces subdiffusion by altering the timing between two diffusional steps yielding a diffusion equation. The mechanism underlying subdiffusion pattern remains not being resolved and coexistence of two different processes such as diffusion on fractals and CTRW are the subject of research in recent years [40, 41].

Using percolation cluster, fBm, CTRW and the synergy of CTRW and percolation cluster and with help of extensive simulations we sample the times an enzyme needs to locate its target, a specific sequence on an invading stretch of DNA randomly positioned in the cellular cytoplasm (the volume not occupied by the native DNA). The average target knockout time, equivalent to the mean first passage time (MFPT) to hit the target in an active state, is studied as function of non-specific binding constant K_{ns}^0 of active EcoRV to DNA and the bond occupation probability p or anomaly in fBm and CTRW. Our results show, however, under the assumption of normal diffusion in the cell the performance of EcoRV is only marginally better than that of a 100%-active mutant: normal diffusion on the length scales of an *E.coli* cell provides very efficient mixing, and the reduced activity of EcoRV would not constitute an advantage but under subdiffusion, EcoRV's performance is surprisingly high. This provides a concrete solution to the subdiffusion-efficiency paradox and supports current ideas that subdiffusion does not contradict efficient molecular reactions in cells.

Chapter 1

Fractals

The word fractal was published by Benoît Mandelbrot [42], who is often called the 'father of fractals' in 1975. It was derived from the Latin fractus meaning "broken" or "fractured". The idea of fractals goes back to the 17th century, when mathematician and philosopher Gottfried Leibniz, as described in his work "The Monadology"[43], considered recursive self-similarity. In 1872 Karl Weierstrass introduced a function with the non-intuitive property of being everywhere continuous but nowhere differentiable and its graph is a fractal in today's language. In 1904, Helge von Koch gave a more geometric definition of a similar function, which is now called the Koch curve [44]. Jean Baptiste Perrin, groping towards a fractal description of a natural phenomenon, discussed the indeterminate nature of flocculated soap flake precipitate out of soap solution [45, 46]. In 1915, Waclaw Sierpinski constructed his triangle and one year later, his carpet. Georg Cantor also gave examples of subsets of the real line with unusual properties. These Cantor sets are also now recognized as fractals. In the late 19th and early 20th centuries, Henri Poincaré, Felix Klein, Pierre Fatou and Gaston Julia investigated iterated functions in the complex plane, although they could not visualize the beauty of many of the objects that they had discovered, because modern computer graphics was not invented yet. In the 1960s, Mandelbrot started investigating self-similarity in papers such as *How Long Is the Coast of Britain? Statistical Self-Similarity and Fractional Dimension* [47]. He illustrated this mathematical definition with computer-constructed visualizations. These images captured the popular imagination; many of them were based on recursion, leading to the popular meaning of the term "fractal" [48]. A fractal as Mandelbrot has explained it in his book, is a fragmented geometric shape that can be split into parts, each of which is a reduced-size copy of the whole [49], a property called self-similarity.

Fractals model disorder in nature more successfully than do objects of classical geometry. In the famous words of Mandelbrot, *clouds are not spheres, mountains are not cones, coastlines are not circles, bark is not smooth, nor does lightning travel in the straight line* [49].

Approximate fractals are easily found in nature. These objects display self-similar structure over an extended, but finite, scale range. Examples include clouds, river networks, mountain ranges, snow flakes [50], lightning, cauliflower, and systems of blood vessels, DNA and polymers [51] can be analyzed as fractals. Even coastlines may be considered fractal in nature. The most basic properties of fractals are self-similarity, symmetry under dilation or scaling and the fractal dimension.

We mainly have two kinds of fractals, deterministic fractals and random fractals. Deterministic fractals are geometrical structures with the property that parts of the structure are similar to the whole [30]. The Koch curve and Sierpinski gasket are the well known deterministic fractals and best examples for deterministic fractals.

The Koch curve is constructed from a unit segment. The middle third section is replaced by two other segments of length 1/3, making an equilateral triangle or a tent, as shown in Fig. 1.1a. The same procedure is repeated for each of the four resulting segments (of length 1/3). This process is iterated for infinite number of times. The length of the intermediate curve at the nth iteration of the construction is $(4/3)^n$, where $n = 0$ denotes the original straight line segment. Moreover, the length of the curve between any two points on the curve is also infinite since there is a copy of the Koch curve between any two points. Therefore the length of the Koch curve is infinite, but it is confined to a finite region of the plane. Thus, the Koch curve is somewhat 'denser' than the regular curve of dimension $d = 1$, but certainly not as dense as a two dimensional object. Therefore its dimension should be between one and two. If a regular object - such as a line segment, or a cube- of dimension d is magnified by a factor b, the original object would fit b^d times in the magnified ones. This consideration may serve as a working definition of the fractal dimension, d_f. Formally let $r(S)$ be similarity transformation that maps all points x onto new points $x' = rx$. The set S is called self-similar with respect to the scaling ratio $r < 1$ if S is equal to the union of $n(r)$ replicas of $r(S)$. If this is the case, one may further define the self similarity dimension, D_s [30]:

$$D_s = \frac{\ln n(r)}{\ln(1/r)} \qquad (1.1)$$

In the Koch curve the line is magnified by the factor of three, and there fit exactly four of the original curves. Therefore the Koch curve is self similar with $r = \frac{1}{3}$, $n(r) = 4$, and $d_f = D_s = \ln 4/\ln 3 \sim 1.262$.

The Sierpinski gasket (Fig. 1.1b) is one of the most popular fractals. A geometric method of creating the gasket is to start with a triangle and cut out the middle piece. This results in three smaller triangles to which the process is continued. The nine resulting smaller triangles are cut in the same way, and so on, indefinitely. The gasket is perfectly self similar, an attribute of many fractal images. Any triangular portion is an exact replica of the whole gasket. The resulting fractal dimension is given by $2^{d_f} = 3$, or $d_f = \ln 3/\ln 2 \sim 1.585$. All deterministic fractal lattices are obtained in a similar way to the examples above and all have a generator, it proceeds with a set of operations that are repeated in a recursive way. There are two kinds of generators for deterministic fractals. In one case, the initiator is replaced by smaller

1.1 Properties of fractals

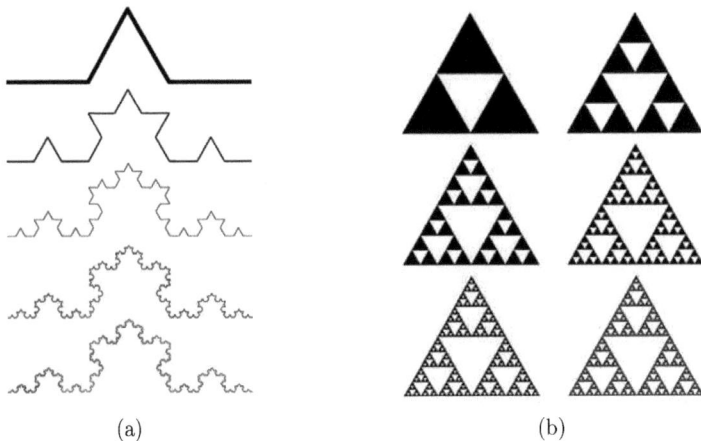

(a) (b)

Figure 1.1: (a) The Koch curve. The initiator is a unit segment. The middle section is replaced by two (similar) sections, forming a tent shape and by doing it again it forms a snowman. After few iterations the last curve is produced. (b) The Sierpinski gasket after few iterations generated from the outside inwards.

replicas of itself and the resulting fractal has then an upper cutoff length which is the length of the initiator. In the alternative approach, replicas of the initiator are assembled into a larger object. In the first case the fractals grow inwards and in the second model grow outwards. The lattice then has a lower cutoff length, but no characteristic large length scale. An ideal fractal lattice has no cut off lengths. But the real-life objects, or fractals constructed in a computer, have both upper and lower cutoffs that represent the size of the fractal structure and the size of its elementary units, respectively [30]. This behavior is typical for natural fractals like biological cell boundaries [52].

1.1 Properties of fractals

The most important property of fractals is their self-similarity, or their symmetry under dilation. To explain what self-similarity means; if we look at a fern leaf (Fig. 1.2), we will notice that every little leaf - part of the bigger one - is very similar to the whole fern leaf. We can say that the fern leaf is self-similar. The same is with fractals: you can magnify them many times and after every step you will see the same shape, which is characteristic of that particular fractal. If we examine the Koch curve we notice that there is central object in the figure which is similar to a snowman. To its right and left there are two other snowmen, each being an exact

Figure 1.2: A self similar fern leaf. Every little leaf is very similar to the whole fern leaf.

reproduction of the central snowman, only smaller by a factor of 1/3. Both of these snowmen display even smaller replicas of themselves to their right and left, etc. In fact, if we look at the Koch curve at any given magnification we will see the same motif again and again.

The main difference between regular Euclidean space and fractal geometries is in symmetry in translation. Regular space is symmetric under translation but in fractals this symmetry is violated. Fractals are symmetric under dilation. Fractals are used for modeling in biology, geography, astrophysics etc. In biology plants and animals exhibit properties of self-similarity as has already been seen in the case of the fern. In humans branches of arteries and veins can be modeled using fractals, as well as kidney structure, heart and brain waves, lungs and the nervous system. In some instances the stock market and economic meters can exhibit properties of self-similarity and as such fractals can be used here.

The other important properties of fractals is their fractal dimension d_f that we explained before.

1.2 Random fractals

Fractals do not need to be generated by deterministic rules. Random fractals are generated by stochastic processes, trajectories of the Brownian motion, Lévy flights, percolation clusters, and diffusion-limited aggregation are good examples of random fractals.

Natural objects do not contain identical scaled down copies within themselves and are more similar to random fractals rather than deterministic fractals. Consider the generation of a random Koch curve, but in the following steps ($k = 2$ onwards), the fractals are obtained by replacing each line segment with the generator in such a way that the triangle of the generator points randomly to either side of the original line. The figure for the final fractal shape looks very irregular compared to the exact Koch curve but is closer to the shape of natural objects such as coastlines. Just as for exact fractals, one can define a fractal dimension for random fractals. One can use the box counting or sand box algorithms to determine the fractal dimension of the random fractals. In the box counting method, space is divided into equal sized cubes (or squares if the figure lies on a plane) of linear dimension r [30]. One then counts the number of cells, $N(r)$ that are needed to cover the given shape. If

$$N(r) \propto r^{-d_f}. \qquad (1.2)$$

as the length r is changed, one says that the distribution of points is d_f-dimensional.

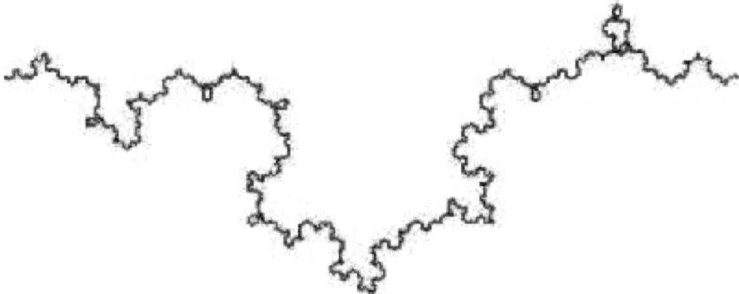

Figure 1.3: Random Koch curve. The fractals are obtained by replacing each line segment with the generator in such a way that the triangle of the generator points randomly to either side of the original line and the final fractal is more similar to the objects in nature like coastlines.

This definition obviously agrees with the Euclidean dimension for straight lines and planes but gives fractional values for more complicated shapes such as coastlines. Note that the equation above is of the same form as that which comes from the definition of the self-similarity dimension mentioned above.

Another example can be Sierpinski carpet (Fig. 1.4a). It is obtained from a unit square initiator. The generator divides this square into nine smaller cells and discards the central cell. In Fig. 1.4a we show the result obtained when the discarded cell is one of the nine cells, chosen at random. Clearly the two figures are related, but the object in Fig. 1.4b is no longer self-similar, instead, we can argue that it is

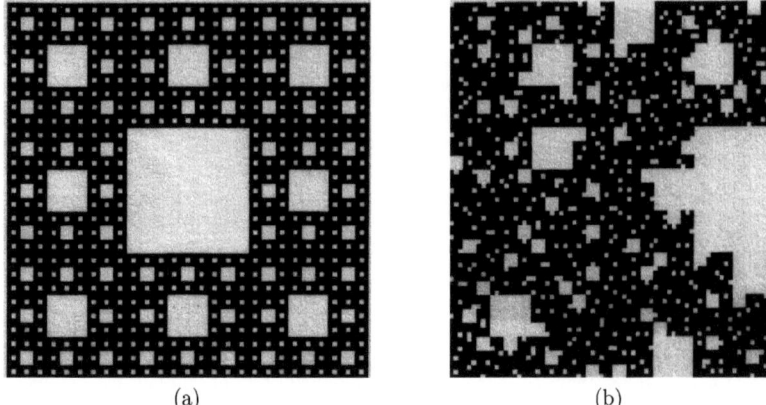

Figure 1.4: (a) The deterministic Sierpinski carpet, generated by removing central subunits of a subdivided squared to 3 by 3 (b) Random Sierpinski carpet, more similar to a sponge.

self-similar in a statistical sense; the distribution of holes looks similar at all length scales. Also on average, the "mass" of the object (the black areas in the figure) increases by a factor of eight when space is dilated by a factor of three. The random carpet has the same fractal dimension $d_f = \ln 8/\ln 3$ as the deterministic carpet. Generally, the mass M of random fractals scales upon dilation, by a factor b, as Euler's homogeneous function

$$M(bL) = b^{d_f} M(L), \quad (1.3)$$

exactly as for deterministic fractals. Note that the solution of this functional equation is

$$M(L) = AL^{d_f}, \quad (1.4)$$

where A is constant.

Indeed, the fractal of Fig. 1.4b resembles a surface of a real sponge more closely than does the original Sierpinski carpet. A similar adaption of the Koch curve, say, may provide an appropriate description of the coastline of Norway. Therefore random fractals are useful as models for natural phenomena. Also in Mandelbrot works, we now know that natural objects are more likely to be fractals rather than not. Actually it is claimed that fractals are more stable and they are connected to self-organized criticality [53].

1.3 Percolation

Percolation is one of the most important and best understood phenomena giving rise to random fractals. We mentioned that many objects in nature resemble a random fractal rather than perfect geometrical shapes. Forming random fractal could be due to dynamic chaotic processes, self organized criticality, etc. Percolation is one of the simplest models for disordered systems [54, 55]. Percolation has lots of applications to such diverse problems as supercooled water, galactic structures, fragmentation porous materials, earthquakes and oil recovery. The word percolation means to cause (a liquid) to pass through a porous body and is taken from coffee percolators. The importance of percolation lies in the fact that it models critical phase transition of rich physical content. Also it can be formulated and understood in terms of very simple geometrical concepts. Historically percolation goes back to Flory and Stockmayer who wanted to describe how small branching molecules form larger macromolecules if more chemical bonds are formed between the original molecules. Percolation processes were introduced in the mathematics literature, in a publication, in 1957 by Broadbent and Hammersley. Hammersely had called the new computer one of the reasons of developing the percolation theory that even up today, they play a very crucial role for percolation in lattices that have thousands of millions of sites being simulated [55].

First we explain percolation theory then discuss its structural properties and different exponents in particular close to the percolation threshold.

1.3.1 Percolation transition

Consider a square lattice on which each bond is present with probability p, or absent with probability $1-p$. Present or empty bonds can stand for different physical properties. It can be pathway for electrical conductors and absent bonds can represent insulators and the electrical current can flow only thorough present bonds. When p is small, there is dilute population of bonds, and the cluster of small numbers of connected bonds predominate. Therefore at low p, the mixture is still an insulator for the electrical current, since a conducting path that connects opposite edges does not exist [55]. As p increases the size of the clusters also increase. Eventually, for p large enough, there exists a cluster that spans the lattice from edge to edge (Fig. 1.5) and we have electrical current through the cluster. If the lattice is infinite, the inception of spanning cluster occurs sharply upon crossing a critical threshold of the bond concentration, $p = p_c$. The probability that a given bond belongs to the incipient cluster, P_∞ undergoes a phase transition: it is zero for $p < p_c$, and increases continuously as p is made larger than the critical threshold p_c (Fig. 1.6) Above and close to the transition point, P_∞ follows a power law:

$$P_\infty \sim (p - p_c)^\beta \quad (1.5)$$

Figure 1.5: Bond percolation on the square lattice. Bonds are present on 25 by 25 square lattices with probability (a) $p = 0.3$ and (b) $p = 0.4$ (c) $p = p_c = 0.5$. In (a) and (b) when $p < p_c$ the clusters are growing by increasing p but still there is no spanning cluster in (c) $p = p_c$ the cluster spanning the lattice appears for the first time

This phenomenon is known as percolation transition. P_∞ plays the role of an order parameter, analogous to magnetization in a ferromagnet, and β is the critical exponent of the order parameter. At the transition point the electrical current can percolate through the medium for the first time. In fact, the transition is similar to all other continuous (second-order) phase transitions in physical systems. In the

1.3 Percolation

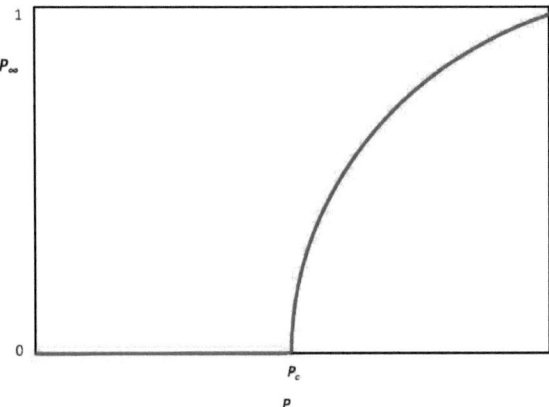

Figure 1.6: Percolation transition. The probability $P(p)$ that a bond belong to the spanning cluster undergoes a sharp transition. Below the critical probability threshold p_c, there is no spanning cluster so $P(p)$ is zero but $P(p)$ becomes finite as soon as $p = p_c$.

Figure 1.7: (a) Site percolation in square lattice above criticality (b) Continuum percolation of circles on the plane. As the concentration of the circles increases the clusters grow in size till the spanning cluster appears.

conductor case p_c would separate the conducting and non-conducting phases.

There exist other types of percolation models. In site percolation the percolating elements are lattice sites, rather than bonds. In that case we think of nearest - neighbor sites as belonging to the same cluster (Fig. 1.7a). Continuum percolation is defined without resorting to a lattice - consider for example a set of circles randomly placed on a plane, where contact is made through their partial overlap (Fig. 1.7b). Finally one may consider percolation in different space dimensions. The percolation threshold p_c is affected by these various choices (table 1.1), but critical exponents such as β, depend only upon the space dimension.

One example of bond percolation in biology is the spreading of an epidemic. The epidemic starts with a sick individual which with probability p can infect its nearest neighbors in one time step and dies after that. In this case the critical concentration p_c separates the phase in which epidemic always dies after finite number of time steps from the phase where epidemic continues forever.

Let us define some more of these important critical exponents. The typical length of finite clusters is characterized by the correlation length ξ. It diverges as p approaches p_c as

$$\xi \sim |p - p_c|^{-\nu}, \tag{1.6}$$

with the same critical exponent ν below and above the transition. The average mass (the number of sites in site percolation, or number of bonds in bond percolation) of finite clusters, S, is analogous to the magnetic susceptibility in ferromagnetic phase transitions. It diverges about p_c as

$$S \sim |p - p_c|^{-\gamma} \tag{1.7}$$

again with the same exponent γ on both sides of transition.

In table 1.1 the value of p_c for different lattices in different dimension is given. In Bond percolation the critical concentration is always smaller than p_c for site percolation, since a bond has always more neighbors than a site. For example in a square lattice a bond has six nearest neighbors and in site percolation case a site has four nearest neighbor. Thus large clusters of bonds can be formed faster than clusters of sites and lower concentration of bonds is needed to form the spanning cluster.

In this chapter we described percolation transition and its different quantities. These quantities are characterized by critical exponents σ and τ, γ, ν and β which are not independent from each other. If one knows two exponent the others follow. For length scales smaller than the correlation length, finite cluster or infinite cluster is self-similar and can be characterized by fractal dimension d_f and graph or chemical dimension d_l. Above p_c for length scales longer than ξ, the infinite cluster is homogeneous and has the dimension d of the lattice.

Lattice	Site Percolation	Bond percolation
Square	0.592746	0.5
Traingular	0.5	$2\sin(\pi/18)$
Simple cubic lattice (first nearest neighbor)	0.311605	0.2488126
Simple cubic lattice (Second nearest neighbor)	0.137	-
Continuum percolation $d = 2$	0.312 ± 0.005	-
Continuum percolation $d = 3$	0.2895 ± 0.0005	-

Table 1.1: Percolation threshold for several two and three dimensional lattices

1.3.2 Structural properties of the percolation

In 1977 Stanley [56] showed that the structure of percolation cluster can be described by fractal concept. Consider first, the infinite cluster at the critical threshold p_c. An example of the infinite cluster is shown in Fig. 1.8. The cluster contains holes

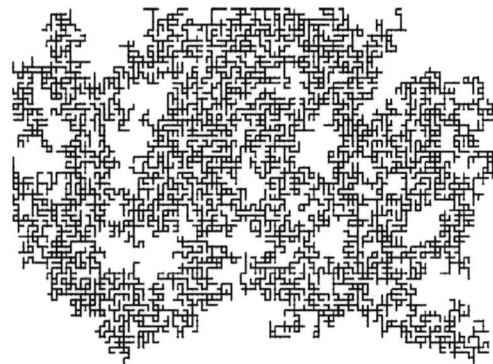

Figure 1.8: Infinite cluster percolation connects all side of the lattice together.

on all length scales, similar to the random Sierpinski carpet of Fig. 1.4b and is self similar in all length scales. Therefore it can be considered as a fractal and it can be proved by a box counting algorithm too. The fractal dimension d_f describes how on the average the mass S within a sphere of radius r scales with r:

$$S(r) \sim r^{d_f} \qquad (1.8)$$

We are dealing with random fractals, therefore, in order to have exact results $S(r)$ is obtained by averaging over many cluster configurations (in different percolation simulations), or, equivalently, averaging over different positions of the center of the sphere in a same infinite cluster.

1.3 Percolation

Below the percolation threshold the mean size of finite clusters, is of the order of the correlation length ξ. Therefore, clusters below criticality can be self-similar only up to length scale ξ. Above criticality, ξ is a measure of the size of finite clusters in the system. The infinite cluster remains infinite in extent, but its largest holes are also of size ξ. It means that the infinite cluster can be self-similar only up to length scale ξ. At distances larger than ξ self-similarity is lost and the infinite cluster

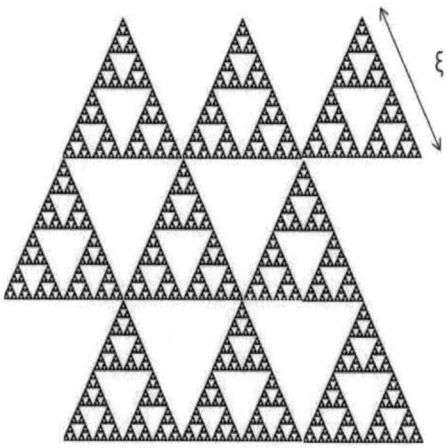

Figure 1.9: The infinite percolation cluster at criticality are presented by Sierpinski gasket. There is self-similarity for distances shorter than ξ, and for larger length scales the cluster is homogeneous.

becomes homogeneous. In other words, for length scales shorter than ξ the system self similar and we see a fractal structure whereas for length scales larger than ξ the system is homogenous. The situation is sketched in Fig. 1.9, in which the infinite cluster above criticality is likened to a regular lattice of Sierpinski gaskets of unit cell size ξ. This can be mathematically summarized as:

$$S(r) = \begin{cases} r^{d_f} & r < \xi, \\ r^d & r > \xi. \end{cases} \quad (1.9)$$

The probability that an arbitrary site, in a region of volume V, belongs to the infinite cluster is S/V. That is the ratio between the number of sites on the finite cluster and total number of sites. If the linear size of the region is smaller than ξ, the cluster is self-similar, so

d	2	3	4	5	6
d_f	91/48	2.53± 0.02	3.05± 0.05	3.69± 0.02	4
d_{min}	1.1307± 0.0004	1.374± 0.004	1.60 ± 0.05	1.799	2
d_{red}	3/4	1.143 ± 0.01	1.385 ± 0.055	1.75 ± 0.01	2
d_f^{BB}	1.6432 ± 0.0008	1.87 ± 0.03	1.9 ± 0.2	1.93 ± 0.16	1/2
ν	187/91	2.186 ±0.002	2.31±0.02	2.355 ± 0.007	5/2

Table 1.2: Fractal dimensions of the substructures of percolation clusters [30].

$$P_\infty \sim \frac{r^{d_f}}{r^d} \sim \frac{\xi^{d_f}}{\xi^d}, \quad r < \xi. \tag{1.10}$$

using Eqs.(1.5) and (1.6) we can express both sides of Eq.(1.10) as power of $p - p_c$:

$$(p - p_c)^\beta \sim (p - p_c)^{-\nu(d_f - d)}, \tag{1.11}$$

and

$$d_f = d - \beta/\nu. \tag{1.12}$$

Therefore the fractal dimension of percolation is not a new, independent exponent, but depends on the critical exponents β and ν. Since β and ν are universal, d_f is also universal.

The fractal dimension is not sufficient to fully characterize the geometrical properties of percolation clusters. Different geometrical properties are important according to the physical application of the percolation model. For example we consider the shortest path between two sites on the cluster. We call this path the *chemical distance* l. The chemical dimension or topological dimension describes how the mass of the cluster within a chemical length l scales with l while the fractal dimension d_f describes how the mass of the cluster scales with the *Euclidean* distance r:

$$S(l) \sim l^{d_l}. \tag{1.13}$$

By comparing Eqs.(1.8) and (1.10), one can infer the relation between regular Euclidean distance and chemical distance:

$$r \sim l^{d_l/d_f} \equiv l^{\nu_l}. \tag{1.14}$$

This relation is often written as $l \sim r^{d_{min}}$, where $d_{min} \equiv 1/\nu_l$ can be regarded as the fractal dimension of the minimal path.

The concept of the chemical length finds several interesting applications, such as oil recovery, in which the first-passage time from the injection well to a production well in a distance r, is related to l. It is also useful in the description of propagation of epidemics and forest fires. Suppose that trees in a forest are distributed as the percolation model. Assume that fire distributes from a burning tree to a tree close to it in each unit time. The fire front will then advance one chemical shell (sites at equal chemical distance from a common origin) per unit time. The speed of propagation would be [30]

$$v = \frac{dr}{dt} = \frac{dl}{dt} \sim l^{\nu_l - 1} \sim (p - p_c)^{\nu(d_{min} - 1)}. \tag{1.15}$$

In $d = 2$ the exponent $\nu(d_{min} - 1) \approx 0.16$ is small and so the increase of v upon crossing of p_c is steep and a fire that could not propagate at all below p_c, may propagate very fast when the concentration of trees is only slightly bigger (just above pc).

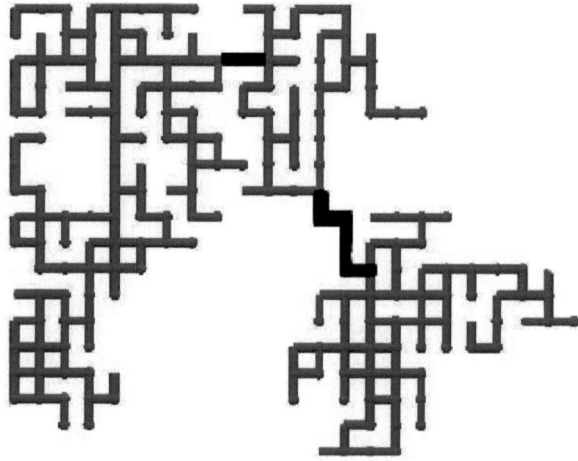

Figure 1.10: Bold solid lines are red bonds which carry the whole current and deleting them stops the current flow.

Critical exponents such as ν, β are universal and depend only on the dimensionality of space, not on other details of the percolation model. d_f and d_l are not the only exponents characterizing percolation cluster at p_c. A percolation cluster is composed of several fractal sub-structures which can be described by other components. Suppose that one applies a voltage on two sites of a metallic percolation cluster. The *backbone* of the cluster consists of those bonds (or sites) which carry the electric current. The *red bonds* carry the total current and deleting a red bond stops the current flow (Fig. 1.10). In table 1.2 we list the values of some of the percolation exponents discussed above.

The fractal dimension of the backbone, d_f^{BB}, is smaller than the fractal dimension of the cluster (see Table. 1.2). That means, most of the mass of the percolation cluster is concentrated in the dangling ends, and the fractal dimension of dangling ends is equal to the fractal dimension of infinite cluster. The fractal dimension of the backbone is known only from numerical simulation.

Chapter 2

Diffusion

Here we discuss dynamical properties of percolation systems. We will show that due to the fractal nature and self-similarity of percolation systems physical laws of transport are changed and are *anomalous*. We consider the total resistance and conductivity and the mean squared displacement and the probability density of random walks as two representative examples. To see the anomalous behavior on fractals and percolation clusters, we first discuss transport and random walk in regular lattices. The problem of random walk in a network or lattice was first studied by G. Pólya in 1919 [57]. That is a stochastic process of a wandering point moving between the sites (nearest neighbors) of the simple cubic lattice and the position changes by ±1 just along one of the axes in any step. Random walk represents the thermal motion of electrons in a metal or moving the holes in semiconductor. It can describe Brownian motion of a particle, the spreading of a drop of ink in a glass of water or bacterial motion.

2.1 Random walks

The simplest example of a *random walk* is a stochastic process of a wandering point moving between the sites (the nearest neighbors), with equal probabilities, of the simple hypercubic lattice and the random walker moves just along one of the coordinate axes in any step. Usually the time variable is considered to be discrete. Due to moving randomly to the new position and choosing the new site independently from the history of the walk, a random walk process is Markovian. If $\mathbf{e}_i$ is a unit vector pointing to a nearest-neighbor site, after n steps, the displacement is:

$$\mathbf{r}(n) = \sum_{i=1}^{n} \mathbf{e}_i, \qquad (2.1)$$

2.1 Random walks — Diffusion

Hence the average displacement will be zero because $\langle e_i \rangle = 0$. Since $\langle e_i.e_j \rangle = 0$ for $i \neq j$ and $\langle e_i.e_i \rangle = 1$, the mean squared displacement is:

$$\langle r^2(n) \rangle = \left\langle \left(\sum_{i=1}^{n} \mathbf{e}_i \right)^2 \right\rangle = n. \tag{2.2}$$

Since the step time unit is τ, $t = n\tau$, then

$$\langle r^2(n) \rangle = (2d)Dt. \tag{2.3}$$

If a is the lattice spacing then $D = a^2/2d\tau$ is the *diffusion constant* and governs how quickly the random walker will spread out into the environment. This kind of motion is known as normal or regular diffusion. The mean squared displacement is the second moment of probability density $P(\mathbf{r}, t)$, the probability that the walker has displaced to r after time t. We can calculate the mean squared displacement by solving $\langle r^2(n) \rangle = \int r^2 P(\mathbf{r}, t) d^d\mathbf{r}$. $P(\mathbf{r}, t)$ can be calculated in one dimension easily. Assume that the jumps in one direction occur with equal probability $p = 1/2$, the probability to reach site m after N step is given by binomial distribution

$$p(m, M) = \left(\frac{1}{2} \right)^N \frac{N!}{\left(\frac{N+m}{2} \right)! \left(\frac{N-m}{2} \right)!}. \tag{2.4}$$

using the Sterling approximation we have

$$\log N! \cong \left(N + \frac{1}{2} \right) \ln N - N + \frac{1}{2} \ln 2\pi, \tag{2.5}$$

and from the expansion $\ln(1+z) \sim z - z^2/2$, when $z \ll 1$ one obtains

$$\ln P(x, N) \approx -\frac{1}{2} \ln N + \ln 2 - \frac{1}{2} \ln 2\pi - \frac{m^2}{2N}. \tag{2.6}$$

Then

$$P(m, N) \approx \sqrt{\frac{2}{\pi N}} e^{-\frac{m^2}{2N}}. \tag{2.7}$$

We define

$$P(x, N) \Delta x = P(m, N) \Delta m. \tag{2.8}$$

Then

$$P(x, N) = \frac{\Delta m}{\Delta x} \sqrt{\frac{2}{\pi N a^2}} e^{-\frac{x^2}{2Na^2}}. \tag{2.9}$$

We know $\Delta x = 2a\Delta m$, therefore we have a Gaussian distribution

$$P(x, N) = \sqrt{\frac{1}{2\pi N a^2}} e^{-\frac{x^2}{2Na^2}}. \tag{2.10}$$

An interesting feature of regular diffusion is that its time dependence is universal, regardless of the dimension of the space, d.

2.2 The probability density

It will be interesting to find a more general derivation of the probability density of being at $\mathbf{r}$ after n steps. We now tackle this question, following the method of characteristic functions. Consider a walk that takes place in $\mathbf{R}^d$ (continuous space) and the steps are drawn from the probability density $p(\mathbf{r}')$, and assume that the walk starts at the origin. The simple random walk is a special case, in which $p(\mathbf{r}') = [1/(2d)] \sum_i \delta(r' - ae_i)$.

The Fourier transform of P is called the *characteristic function* of the probability density:

$$F_n(\mathbf{k}) = \int P_n(r) e^{i\mathbf{k}\cdot\mathbf{r}} d^d r \qquad (2.11)$$

Because the steps are independent, the process obeys Markov property, then

$$P_{n+1}(\mathbf{r}) = \int P_n(\mathbf{r}') p(\mathbf{r} - \mathbf{r}') d^d r' \qquad (2.12)$$

Therefore, the Fourier transform is:

$$F_{n+1}(\mathbf{k}) = F_n(\mathbf{k})\lambda(\mathbf{k}), \qquad (2.13)$$

Where $\lambda(k)$ is the characteristic function of step probability density. It is also known as step structure function. It is defined as:

$$\lambda(k) = \int p(r') e^{i\mathbf{k}\cdot\mathbf{r}'} d^d r'. \qquad (2.14)$$

A recursive relation from Eq.(2.13) yields:

$$F_n(\mathbf{k}) = \lambda(\mathbf{k})^n. \qquad (2.15)$$

We obtain the probability density from the inverse transform

$$P_n(\mathbf{r}) = \frac{1}{(2\pi)^d} \int F_n(\mathbf{k}) e^{-i\mathbf{k}\cdot\mathbf{r}} d^d k. \qquad (2.16)$$

Suppose that the step probability function has zero mean and finite variance, then at long times the distribution $P_n(\mathbf{r})$ will be a Gaussian. This is a result of the *central limit theorem* that states conditions under which the mean of a large number of independent random variables, each with finite mean and variance, will be approximately normally distributed [58, 59]. In simple one dimensional case, according to Eq.(2.14) and knowing that $\int \mathbf{r} p(\mathbf{r}) dr = 0$ and $\int r^2 p(\mathbf{r}) dr = a^2$, the structure function is,

$$\lambda(k) = 1 - \frac{1}{2}k^2 a^2 + O(k^2), \qquad (2.17)$$

2.2 The probability density

and therefore the characteristic function is

$$F_n(k) = e^{-nk^2 a^2/2 + nO(k^2)}. \tag{2.18}$$

For very long times when $(n \gg 1)$, the main contribution to $F_n(k)$ comes from $|k| < 1/n^{1/2}$, and $O(k^2) \sim 1/n^{1/2}$ can be neglected. Then from the inverse transform, we have

$$P_n(r) = \frac{1}{2\pi} \int_{-\infty}^{+\infty} e^{(-nk^2 a^2/2 - ik \cdot r)} dk = \frac{1}{\sqrt{2\pi a^2 n}} e^{-r^2/(2a^2 n)}. \tag{2.19}$$

or in terms of time variable $t = n\tau$

$$P(r,t) = \frac{1}{\sqrt{4\pi D t}} e^{-r^2/(4Dt)}. \tag{2.20}$$

Which is the same result as Eq.(2.6) since $D = 1/2$ in our one dimensional case [30]. The probability density in d dimensional space is essentially the same, and just the normalization factor should change to $(4\pi Dt)^{-d/2}$.

Diffusion is known as the limit of continuous time and continuous space. Let us consider a discrete time and space random walk in one dimension. At each time step the walker hops with the equal probability $\frac{1}{2}$ to the nearest site to its right or left. So the probability of being at site m at the nth step, $P_n(m)$ satisfies the Eq.(2.3):

$$P_{n+1}(m) = \frac{1}{2} P_n(m-1) + \frac{1}{2} P_n(m+1),$$

or

$$P_{n+1}(m) - P_n(m) = \frac{1}{2} P_n(m-1) - P_n(m) + \frac{1}{2} P_n(m+1), \tag{2.21}$$

It is easier to deal with differential equations and this motivates the passage to the continuum limit. We introduce the variables $x = ma$ and $t = n\tau$, and take the limit $a \to 0$ and $\tau \to 0$. Then the probability $P_n(m)$ will be replaced by probability density $P(x,t)$. We keep $a^2/2\tau = D$ constant in the limiting process and Eq.(2.21) becomes

$$\frac{\partial}{\partial t} P(x,t) = D \frac{\partial^2}{\partial x^2} P(x,t). \tag{2.22}$$

This is the diffusion equation in one dimension and with initial condition $P(x,0) = \delta(x)$ the solution will be Eq.(2.20). Indeed in the long time limit there is no difference between discrete random walks and diffusion. In higher dimensions the diffusion equation is

$$\frac{\partial}{\partial t} P(r,t) = D \nabla^2 P(r,t), \tag{2.23}$$

where ∇^2 is the d-dimensional Laplacian operator.

We had assumed the in one dimensional walk the random walker moves to its right or left with the equal probability. Let us assume the probabilities to move to right and left are not equal and are r and l such as $r + l \leq 1$. This causes a drift to the right. Therefore the diffusion equation changes to

$$\frac{\partial}{\partial t}P(x,t) = -v\frac{\partial}{\partial x}P(x,t) + D\frac{\partial^2}{\partial x^2}P(x,t), \qquad (2.24)$$

Eq.(2.24) is a Fokker-Planck equation where

$$D = \lim_{\substack{\Delta x \to 0 \\ \Delta t \to 0}} \frac{k\langle(\Delta x)^2\rangle}{2\Delta t}, \quad v = \lim_{\substack{\epsilon \to 0 \\ \Delta x \to 0 \\ \Delta t \to 0}} \frac{\Delta x \epsilon}{\Delta t} \qquad (2.25)$$

that $k = l + r$ and $\epsilon = r - l$. $D = a^2/(2\tau)$ and $v = \epsilon a/\tau$ is called the drift velocity. Eq.(2.24) is known as the classical diffusion equation with drift (where the parameter v governs the speed of drift to either the left or the right). Note that in the convection-diffusion equation, the factor v/D and $1/\Delta x$ diverge as Δx goes to zero in continuum limit. Therefore the convective term $\partial P/\partial t$ dominates over the diffusion term $\partial^2 P/\partial t^2$. Then ϵ must be proportional to Δx, when $\Delta x \to 0$ to have a nonpathological continuum limit and ensure that both the first and the second order spatial derivative are finite [58].

If $r = l$, then $v = 0$ and there is no drift (and no drift term in Eq.(2.24)). In this case, we end up with the classical diffusion equation which is the governing equation of the simple random walk or a Brownian motion. The parameter D is the diffusion coefficient. If $r + l < 1$, then we can have resting phases or waiting times in the random walk. This gives k (and D) a value than the case $r + l = 1$. This leads to an intuitive interpretation that random walkers who occasionally stop to rest will spread out in the environment slower than those that don't.

We also write the Eq.(2.24) as a continuity equation that expresses the conservation law meaning that the walker cannot be created or destroyed;

$$\frac{\partial}{\partial t}P(x,t) = \frac{\partial}{\partial x}J(x,t), \qquad (2.26)$$

where $J \equiv vP - D\partial P/\partial x$ is the probability current [30].

We find a relation between macroscopic transport parameters like *conductivity* and microscopic coefficient of diffusion if we let the walker to posses an electric charge e, in a metal that is restricted to half of the space $x > 0$ at temperature T. Then there will be an electric field E and a constant velocity v and by applying Ohm law we will have : $nev = -\sigma E$, where n is the density of charges per unit volume. We will have $\partial P/\partial t = 0$ and from Eq. (2.24) we obtain:

$$\frac{\sigma E}{ne}\frac{\partial P}{\partial x} + D\frac{\partial^2 P}{\partial x^2} = 0; \qquad (2.27)$$

solution of this equation is $P(x) = Ce^{-\sigma Ex/(neD)}$ (by applying the boundary conditions and considering that there is no flux in $x < 0$ therefore $J(x = 0) = 0$). On the other hand the charges will be in thermal equilibrium which is characterized by Boltzman distribution $P_{eq} = Ce^{-Eex/(k_BT)}$. By comparing these two results we have:

$$\sigma = \frac{ne^2}{k_BT}D. \tag{2.28}$$

This formula is called Einstein relation and was introduced in 1905 as an unexpected connection between conductivity (macroscopic parameter) and diffusion coefficient (microscopic coefficient).

2.3 Anomalous diffusion

Random walks obey Gaussian statistics, and their mean squared displacement increases linearly in time; $\langle r^2 \rangle \sim t$. In many physical systems, however, it is found that diffusion follows an anomalous pattern: the mean-square displacement is $\langle r^2 \rangle \sim t^{2/d_w}$, where $d_w \neq 2$. Here we shortly introduce different kinds of anomalous diffusion including continuous time random walks (CTRWs) (with heavy tailed waiting time distributions), and a variation of fractional Brownian motion (FBM) model.

2.3.1 Continuous time random walk

We use CTRW model with slow-decaying waiting times between consecutive steps to illustrate a different cause for anomalous diffusion. First let us introduce continuous-time random walk (CTRW) and then we will discuss anomalous CTRW. The continuous time random walk (CTRW) was introduced by Montroll and Weiss [60]. Unlike discrete time random walks, time steps are random variables. We assume that $\psi(t)$ is the probability density of waiting times. Then the probability that the waiting time between steps is greater than t is the cumulative function of ψ between time t and infinity

$$\Psi(t) = \int_t^\infty \psi(t')dt' \tag{2.29}$$

We define $\psi_n(t)$ as the probability density that nth jump happens at time t.

$$\psi_{n+1}(t) = \int_0^t \psi_n(t')\psi(t-t')dt'. \tag{2.30}$$

So once the walker arrives at time $t' < t$ it stays there till time t. We define $\hat{\psi}(s)$ as the Laplace transform of $\psi(t)$

$$\hat{\psi}(s) = \int_0^\infty \psi(t)e^{-st}dt, \tag{2.31}$$

then we will find
$$\hat{\psi}_n(s) = \hat{\psi}^n(s), \quad \hat{\Psi}(s) = \frac{1-\hat{\psi}(s)}{s}. \tag{2.32}$$
We introduce the probability of being at point r at the nth step, $P_n(r)$
$$P(r,t) = \sum_{n=0}^{\infty} P_n(r) \int_0^t \psi_n(t')\psi(t-t')dt' \tag{2.33}$$
By applying Laplace transform we obtain
$$\hat{P}(r,s) = \sum_{n=0}^{\infty} P_n(r)\hat{\psi}(s)\frac{1-\hat{\psi}(s)}{s} \tag{2.34}$$
Fourier transform of Eq.(2.34) gives
$$\hat{P}(k,s) = \frac{1-\hat{\psi}(s)}{s[1-\hat{\psi}(s)\lambda(k)]} \tag{2.35}$$
This equation is called the Montroll-Weiss equation and is an exact solution in Fourier-Laplace space. For the special case $\psi(t) = \delta(t-\tau)$, $\langle \hat{r}^2(t)\rangle = t$, the random walker performs normal diffusion. Now We assume $\psi(t)$ be a power-law waiting time distribution
$$\psi(t) \sim At^{-(\alpha+1)}, \quad 0 < \alpha \le 1 \tag{2.36}$$
which has no characteristic time scale. The divergence of the characteristic waiting time $\int_0^\infty t\psi(t)dt$ causes interesting effects such as aging and weak ergodicity breaking [61, 62, 63]. The mean squared displacement can be found from Eq.(2.35)
$$\langle \hat{r}^2(s)\rangle = (-i)^2 \frac{\partial^2 P(\hat{k},s)}{\partial k^2}|_{k=0} \tag{2.37}$$
For slow decaying $\psi(t) \sim At^{-(\alpha+1)}$ mean-square displacement is
$$\langle \hat{r}^2(t)\rangle \sim t^\alpha, \quad 0 < \alpha \le 1 \tag{2.38}$$
that shows a subdiffusive behavior for CTRW.

2.3.2 Random walks and diffusion on fractal objects

The trail of a random walker is a complicated random object. The trail is self-similar and can be considered as a fractal. The fractal dimension of a random walk is called the walk dimension and is denoted by d_w. If we think of the sites visited by a walker as 'mass', then the mass
$$M \sim t \sim r^{d_w}, \tag{2.39}$$
Where r is the typical distance covered after time t. The mean squared displacement is then given by
$$\langle r^2(t)\rangle \sim t^{2/d_w}. \tag{2.40}$$
For regular diffusion $d_w = 2$, but in fractals $d_w \ne 2$ and the diffusion is anomalous.

2.3 Anomalous diffusion
Diffusion

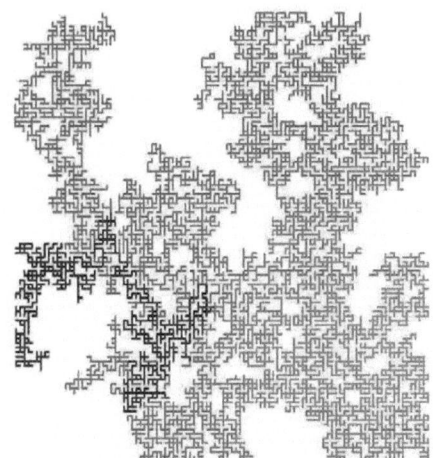

Figure 2.1: Random walk path in the percolation cluster at criticality.

2.3.3 Diffusion in percolation clusters

De Gennes [64] pondered the problem of a random walker in percolation clusters, which he described as "the ant in the labyrinth". We mentioned that percolation clusters may be considered as random fractals. Therefore diffusion in percolation clusters can be explained as diffusion in fractals. Then the mean squared displacement of a random walker diffusing in a percolation cluster is

$$\langle r^2(t) \rangle \sim t^{2/d_w}. \tag{2.41}$$

Remembering that the probability of being on the infinite cluster is proportional to $|p - p_c|^\beta$, we consider the diffusion on percolation clusters in three different cases; above criticality, at criticality and below criticality and study long-time asymptotic limit of a time-independent diffusion constant $D(p,t) = \langle r^2(t) \rangle / t$ [30]:

- Above criticality $(p > p_c)$, $r(t) \ll \xi$ that results in long ranged diffusion coefficient $D \sim (p - p_c)^{\mu - \beta}$. The motion is subdiffusive in short times and in long times it changes to normal diffusion, since the cluster is homogeneous in long scales.

- At criticality $(p = p_c)$, $D \sim t^{(2/d_w)-1}$. The percolation cluster is a fractal and self-similar on all length scales. The motion is subdiffusive in all time scales except in first few steps where the random walker meets no obstacle.

d	d_w	$\bar{\mu} = \mu/\nu$	d_w^{BB}
2	2.878± 0.001	0.9826± 0.0008	2.62± 0.03
3	3.88± 0.03	2.26± 0.04	3.09 ± 0.03
4	4.68 ±0.08	3.63 ± 0.03	-
5	5.50 ± 0.06	4.81 ± 0.04	-
6	6	6	4

Table 2.1: Dynamical exponents for percolation

- Below criticality $(p < p_c)$, when $t \to \infty$, $D \sim t^{-1}(p - p_c)^{-2\nu}$. The random walker is trapped in local environment and the long ranged diffusion coefficient becomes zero.

The scaling function of D that is consistent with these properties, is [30]

$$D(t,p) = t^{2/d_w - 1} f(\epsilon t^{(d_w - 2)/(\mu - \beta)d_w}), \quad (2.42)$$

where $\epsilon = (p - p_c)/p_c$ measures the distance from criticality and

$$f(x) \sim \begin{cases} x^{\mu - \beta} & \text{as } x \to \infty, \\ \text{constant} & \text{as } x \to 0, \\ (-x^{-2\nu}) & \text{as } x \to -\infty. \end{cases} \quad (2.43)$$

consistency yields the relation

$$d_w = 2 + \frac{\mu - \beta}{\nu} \quad (2.44)$$

and $d_f = d - \beta/\nu$.

Numerical estimates of the transport exponents in percolation are listed in Table 2.1.

2.3.4 Spectral dimension and fractons

Spectral (or fracton) dimension is used to describe the dynamic of fractal networks. It is defined by

$$d_s = \frac{2d_f}{d_w}, \quad (2.45)$$

therefore from Eq.(2.40), anomaly α can be defined by the spectral and fractal dimension.

$$\langle r^2(t) \rangle \sim t^\alpha, \quad \alpha = \frac{d_s}{d_f} \quad (2.46)$$

Alexander and Orbach [65] conjectured that the fracton dimension of an infinite percolation cluster is

$$d_s = \frac{4}{3}, \tag{2.47}$$

for all dimensions $d > 1$. Indeed the spectral dimension is exactly $\frac{4}{3}$ in $d \geq 6$, and close to $\frac{4}{3}$ in $2 \leq d < 6$.

Now we consider the vibrational modes of an elastic fractal such as percolation cluster, consisting of particles connected by harmonic springs. In fractals vibrational modes are called fractons rather than phonons. The density of fractons found for fractals $g(\omega) \sim \omega^{d_s-1}$ is valid for the percolation cluster at criticality. Above criticality the crossover behavior in the displacement $\langle r^2(t) \rangle \sim t^{2/d_w}$ ($t < t_{cross} \sim \xi^{d_w}$) and $\langle r^2(t) \rangle \sim t (t > t_{cross})$, gives the characteristic length scale [30]

$$\Lambda(\omega) \sim \begin{cases} \omega^{-2/d_w} & \omega \gg \omega_\xi, \\ \omega^{-1} & \omega \ll \omega_\xi. \end{cases} \tag{2.48}$$

That can be identified as the wavelength of the vibrational modes. In isotropic case in which the spring constants are summed to be scalars one obtains the equations

$$\frac{d^2 U_i(t)}{dt^2} = \sum_j k_{ij} \left[U_j(t) - U_i(t) \right], \tag{2.49}$$

where U_i is the displacement of ith site, and sum runs over all nearest neighbors j of site i. From Eq.(2.49) we obtain the scaling relation $t \sim 1/\omega^2$, so the crossover frequency ω_ξ is

$$\omega_\xi \sim t_{cross}^{-1/2} \sim \xi^{-d_w/2} \sim (p - p_c)^{d_w \nu/2}. \tag{2.50}$$

This frequency separates two vibrational regimes, fractons with a density of states $g(\omega) \sim \omega^{d_s-1}$, and phonons, with density of states $g(\omega) \sim \omega^{d-1}$.

For percolation above criticality the vibrational modes are either localized fractons for $\omega > \omega_\xi$ and their amplitude decays exponentially with distance or phonons, extended over the whole cluster, for $\omega < \omega_\xi$.

2.3.5 Fractional Brownian motion

Another alternative way to study the anomalous diffusion is to generalize the random walk model by adding long range correlations as the fractional Brownian motion (fBm) B^H, which is developed by Mandelbrot [66]. fBm is an extension of the classical Brownian motion that allows its disjoint increments to be correlated with a Hurst parameter $H \in (0, 1)$ that usually starts at zero. Its covariance function is:

$$E[B^H(t) B^H(s)] = \frac{1}{2}(|t|^{2H} + |s|^{2H} - |t-s|^{2H}), \tag{2.51}$$

Covariance function describes the variance of a random variable process or a random field and shows how much two variables change together. The value of parameter H in fBm shows process the fBm is:

- If $H = 1/2$, the process is in fact a Brownian motion or a Wiener process.

- If $H > 1/2$ then the increments of the process are positively correlated. This dependence means that if there is an increasing pattern in the previous "steps", then it is likely that the current step will be increasing (to same direction) as well.

- If $H < 1/2$, the increments of the process are negatively correlated. This dependence means that if there is an increasing pattern in the previous "steps", then it is likely that the current step will be decreasing (to opposite direction).

In stochastic processes we apply fBm as we apply classical Brownian motion $x(t) = x(t-1) + \delta x$. δx is a white noise in classical Brownian motion and in fBm, is a fractional gaussian noise defined as: $\delta x(t) = B^H(t+1) - B^H(t)$. Similar to the normal Brownian motion, fBm is an ergodic process that means it has the same behavior averaged over time as averaged over space and the mean squared displacement is

$$\langle r^2(t) \rangle \sim t^{2H}. \tag{2.52}$$

From Eq.(2.53) we obtain

$$d_w = \frac{1}{H}. \tag{2.53}$$

2.4 First passage time

The first passage underlies many stochastic processes in which the event relies on a variable reaching a specified value for the first time. The problem of determining the first-passage times for diffusion and other Markov processes arises in biological modeling, firing of neurons [67], diffusion-limited aggregation [68], the passage of a biomolecule through a membrane nanopore [69], the encounter of two independently diffusing particles [3] or the electrical current caused by anomalously moving charge carriers in amorphous semiconductors [70] can be mapped onto the problem of calculating the first passage time density (FPTD), and the associated mean first passage time (MFPT) [58].

2.4 First passage time

The First passage probability is the probability that the diffusing particle (or a random walk) first reaches a specified site at a specified time. An important aspect of first passage phenomena is the conditions by which a random walk process terminates. In many cases the diffusing particle physically disappears or dies when the specified point or set of points is hit. A natural way of constructing the FPTD in such a case is the method of images [58]. For instance, on the semi-infinite domain, we use this method and mirror the unrestricted propagator with initial condition x_0.

Another way of obtaining FPTD is applying an absorbing boundary at the target position x_{target} and calculating the negative time derivative of the survival probability. We define $P(\mathbf{r},t)$ as the probability distribution and $F(\mathbf{r},t)$ as the first passage time (FPT) probability. For a random walk to be at $\mathbf{r}$ at time t, the walker must first reach $\mathbf{r}$ at some earlier time t' and the return to $\mathbf{r}$ after $t-t'$ time steps. This connection between $P(\mathbf{r},t)$ and $F(\mathbf{r},t)$ may be expressed as the convection relation, with initial conditions that the walk starts at $\mathbf{r}=0$ [58]

$$P(\mathbf{r},t) = \delta_{\mathbf{r},0}\delta_{t,0} + \sum_{t' \leq t} F(\mathbf{r},t')P(0,t-t'). \qquad (2.54)$$

By Laplace transform we have

$$F(\mathbf{r},s) = \begin{cases} \frac{P(\mathbf{r},s)}{P(0,s)} & \mathbf{r} \neq 0, \\ 1 - \frac{1}{P(0,s)} & \mathbf{r} = 0. \end{cases} \qquad (2.55)$$

Therefore we obtain the first passage time probability from probability distribution. We define survival probability $S(t) = \int_0^\infty P(\mathbf{r},t)d\mathbf{r}$ as the probability of being alive for the walker, then the first passage probability will be $1-S(t)$ and first passage time density, p_{fp} is defined as

$$p_{fp} = -\frac{dS(t)}{dt}. \qquad (2.56)$$

The unbiased Brownian domain with initial condition $P_0(x) = P(x,0) = \delta(x-x_0)$ is described by the FPTD

$$p_{fp}(t) = \frac{x_0}{\sqrt{4\pi D t^3}} e^{\left(\frac{-x_0^2}{4Dt}\right)}, \qquad (2.57)$$

which defines the probability $p_{fp}(t)dt$ for the particle to arrive at $x=0$ during the time interval $t,...,t+dt$. Its long-time behavior corresponds to the $3/2$ power-law behavior

$$p_{fp}(t) \sim \frac{x_0}{\sqrt{D}} t^{-3/2}. \qquad (2.58)$$

In particular, even for Brownian processes there are natural cases when the characteristic time diverges. The mean first passage time (MFPT) $T = \int_0^\infty t p_{fp}(t)dt = \infty$.

In a finite box of size L, the Brownian first passage time problem has an exponential tail, with increasingly quicker decay on increasing mode number,

$$p_L(t) = \frac{\pi}{DL^2} \sum_{n=0}^{\infty} (-1)^n (2n+1) e^{\left(\frac{-D(2n+1)^2 \pi^2 t}{4L^2}\right)}, \qquad (2.59)$$

With an initial condition in the center of a box with two absorbing boundaries, the MFPT becomes $T = L^2/(2D)$.

In CTRW case the essential property of subdiffusive first passage time problems lies in the fact that the long-tailed nature of the waiting time PDF translates into the FPTD itself. The MFPT diverges both in the absence of a bias and under a constant drift, pertaining to both finite as well as semi-infinite domains [71, 72]. For the three cases of first passage time problems, we obtain the following subdiffusive generalizations:

(a) For subdiffusion in the semi-infinite domain with an absorbing wall at the origin and initial condition $P(x,0) = \delta(x - x_0)$, p_{fp} is [71]

$$p_{fp}(t) \sim \frac{x_0}{|\Gamma(-\alpha/2)| K_\alpha^{1/2}} t^{-1-\alpha/2}. \qquad (2.60)$$

The decay becomes a flatter power law than in the Markovian case ($p_{pf}(t) \sim t^{-3/2}$).
(b) p_{fp} for subdiffusion in a finite box is [71]:

$$p_{fp}(t) \sim t^{-1-\alpha}. \qquad (2.61)$$

The exact FPTD of fBm, is not known. It was conjectured [73], based on scaling argument and numerical evidence, that for large t, in semi infinite domain $p_{pf}(t)$ scales with t as

$$p_{fp}(t) \sim t^{H-2}. \qquad (2.62)$$

2.5 Ergodicity

For completeness we mention a subtle point in the analysis of time series of anomalous diffusion processes, namely ergodicity. A system is ergodic if there is at least one trajectory that passes through all points in phase space (for which probability density p is non-zero). The concept of ergodicity is also significant from a measurement perspective because in practical situations, we do not have access to all the sample realizations of a random process. We therefore have to be content in these situations with the time-averages that we obtain from a single realization. The time averaged mean squared of a continuous random process $X(t)$ is defined by

$$\langle x^2 \rangle_T = \frac{1}{2T} \int_0^T X^2(t) dt. \qquad (2.63)$$

The ensemble averaged mean squared displacement is the spatial average over the probability density function $P(x,t)$ to find the particle at position x (in one dimension) at time t.

$$\langle x^2(t) \rangle_N = \int x^2 P(x,t) dx. \qquad (2.64)$$

Then we call a process ergodic in mean squared displacement if the time average asymptotically approaches the ensemble average.

$$\lim_{T \to \infty} \langle x^2 \rangle_T = \langle x^2 \rangle_N \qquad (2.65)$$

2.5.1 Ergodicity in different processes

We consider a process subdiffusive, if the mean squared displacement of processes is [74, 75]

$$\langle x^2(t) \rangle = \frac{2dK_\alpha}{\Gamma(1+\alpha)} t^\alpha, \qquad (2.66)$$

where $0 < \alpha \leq 1$ and K_α with dimension $[K_\alpha] = cm^2 s^{-\alpha}$ is the anomalous diffusion constant. The limit $\alpha = 1$ corresponds to regular Brownian motion. We will see in the next chapter that for regular Brownian motion ensemble averaged mean squared displacement is $\langle x^2(t) \rangle = 2dDt$, is linear in time and also is ergodic.

In order to avoid errors in the ensemble average due to inhomogeneities between individual particles, for instance differences in mass and surface structure, relatively early idea of single particle trajectory analysis was brought forward [76]. From the time series $x(t)$ recorded in such setups that one determines the time averaged mean squared displacement (TA MSD)

$$\overline{\delta^2(\Delta, T)} = \frac{1}{T - \Delta} \int_0^{T-\Delta} (x(t+\Delta) - x(t))^2 dt. \qquad (2.67)$$

Δ is referred to as the lag time and T is the overall measurement time. TA MSD is a two time quantity in the sense that the position difference entering expression (Eq.(2.67)) corresponds to two points of the time series $x(t)$ separated by the lag time Δ. In recent years single particle trajectory analysis has indeed become one of the standard tools to probe the motion of a test particle. This technique reveals the subdiffusive behavior of large molecules in the crowded environment of living biological cells [16, 18, 77] and in reconstituted crowding environment [78, 79]. Subdiffusion was observed on longer scales in the motion of individual bacteria in a biofilm [80]. In many of these studies significant scatter of the amplitude of the TA MSD for different trajectories have been observed [80]. Scattered amplitude between individual trajectories, may have different origins. It may be caused due to spatial inhomogeneities that the tracer particle is in the area with varying diffusivities and the sojourn times in different areas are the impact on the behavior of individual trajectories. Scattered amplitude may be due to dynamic properties of the process (for example in CTRW). Also the fitness of the recorded trajectories could be another reason of the scattered amplitude. There may be no sufficient sampling, in the sense that for the given measurement time, the time series become too short to be statistically sufficient [81].

In the next chapters we study the diffusion of an enzyme inside a cell using different dynamic processes such as bond percolation, CTRW and fBm. In case of diffusion on percolation cluster we see that the ensemble averaged MSD and the TA MSD are equal, therefore the diffusion is ergodic. fBm is a gaussian process with stationary increments and due to its stationary character fBm is ergodic [82] and TA MSD in one dimension is

$$\overline{\delta^2(\Delta, T)} = 2\frac{K_\alpha}{\Gamma(1+\alpha)}\Delta^\alpha, \qquad (2.68)$$

and $\alpha = 2H$ is equivalent to the ensemble average in limit of long measurement time. In the confined geometry both ensemble average and TA MSD reach a plateau determined the interval size [81].

In contrast to fBm and diffusion on percolation cluster, heavy tailed waiting time distributions CTRW causes weak ergodicity breaking. In this case, the ensemble averaged MSD leads to $\langle x^2(t) \rangle = 2K_\alpha t^\alpha / \Gamma(1+\alpha)$ in one dimension. Strong ergodicity breaking is found when a system is divided into inaccessible regions of its phase space. Namely, a particle or a system starting in one region cannot explore all other regions due to some non-passable barrier. In weak ergodicity breaking, the phase space is not divided into inaccessible regions. Instead, due to the power law sticking times, the dynamics are non-stationary and non-ergodic.

In this case in contrast to the ensemble averaged MSD the corresponding ensemble averaged time averaged quantity

$$\langle \overline{\delta^2(\Delta, T)} \rangle = \frac{1}{T-\Delta} \int_0^{T-\Delta} (x(t+\Delta) - x(t))^2 dt, \qquad (2.69)$$

for $\Delta \ll T$ [83, 84] is

$$\langle \overline{\delta^2(\Delta, T)} \rangle \sim \frac{2K_\alpha}{\Gamma(1+\alpha)} \frac{\Delta}{T^{1-\alpha}}. \tag{2.70}$$

This linear lag time dependence might deceivingly indicate normal diffusion. The reason to use the additional ensemble average in Eqs.(2.69) and (2.70) is that the pure time average $\overline{\delta^2(\Delta, T)}$ shows pronounced scatter around its ensemble average Eqs.(2.69). This is caused by the scale free nature of CTRW subdiffusion. In the given time series there may occur a single or few events for which the individual waiting times become of the order of the measurement time T, and no matter how long T is chosen. Therefore in the derivation expression the additional ensemble averaging is necessary to include the average number of steps performed in the time interval Δ [84].

Previously, weak ergodicity breaking was investigated for the following phenomena: blinking quantum dots [85, 86], Lévy walks [87], occupation time statistics of the CTRW model [88], the fractional Fokker-Planck equation [89] and in vivo gene regulation by DNA-binding proteins [22]. Recently, a relation between statistics of weak ergodicity breaking and statistics of non-self averaging in models of quenched disorder was found [90].

Another example of non ergodic process could be synergy of geometrical disorder and energetic disorder [91, 92]. An example is the combination of geometrical restrictions modeled by fractals percolation cluster and of chemical residence times represented by continuous time random walks CTRWs with heavy tails, each leading to subdiffusion. This process is proved to be non ergodic [41]. The ensemble averaged MSD is

$$\langle x^2(t) \rangle \sim t^{\alpha\beta}, \tag{2.71}$$

which α is the anomalous coefficient for CTRW and $\beta = 2/d_w$ is the anomalous diffusion coefficient in percolation. Ensemble averaged time averaged MSD for this case is

$$\langle \overline{\delta^2(\Delta, T)} \rangle \sim t^{1-\alpha+\alpha\beta}. \tag{2.72}$$

Taking $\alpha \to 0$ in Eq.(2.72) leads to a linear time dependence of the time averaged MSD of the subordinated process independent of the underlying fractal structure. On the other hand $\alpha < 1$ leads to a subdiffusive exponent dominated by the fractal nature. Thus, we have a process that recovers time averaged exponents of the MSD close to the fractal ones, yet characterized by ergodicity breaking [41].

To avoid complications in the sense mentioned here, in the following we restrict our discussion to the ensemble quantities such as FPT density, MFPT, etc.

Chapter 3

Diffusion in percolation cluster

3.1 Time dependent first passage properties between two spheres

We would like to obtain the first passage time for a particle (for instance a protein) that diffuses in our model cell, in which we mimic the crowding by allowing the particle to move on a critical bond percolation cluster. As sketched in Fig. 3.1 we impose that the outer sphere with radius R is reflective (cell wall), while the inner radius b defines the nucleoid containing the DNA. A particle arriving to this inner sphere is absorbed, i.e., non-specifically bound to the DNA. First to simplify the problem we solve it for regular space and then we extend it to the fractal space problem.

We use The Green's function technique, to solve the problem in regular space. We solve the problem for the case that the inner sphere is absorbing (at $r = b$) and the outer sphere is reflecting (at $r = R > b$). As the first passage characteristic does not involve angular variables, the systems can be considered one dimensional. Therefore, we take the initial conditions to be a spherical shell of concentration at time $t = 0$, $P(r, t = 0) = \delta(r - r_0)/\Omega_d r_0^{d-1}$ that is initially at $r = r_0$. Ω_d is the normalization coefficient and the surface area of the d-dimensional sphere. In three dimension $\Omega_d = 4\pi$ and in two dimension $\Omega_d = 2\pi$.
The diffusion equation is:

$$\frac{\partial P(r,t)}{\partial t} = D\left[\frac{\partial^2 P(r,t)}{\partial r^2} + \frac{d-1}{r}\frac{\partial P(r,t)}{\partial r}\right], \tag{3.1}$$

The Laplace transform (Eq.(2.31)) of $\frac{\partial^2 P(r,t)}{\partial r^2}$

$$\mathcal{L}\left(\frac{\partial^2 P(r,t)}{\partial r^2}\right) = \frac{\partial^2 P(r,s)}{\partial r^2}, \tag{3.2}$$

3.1 First passage properties between two spheres — Diffusion in percolation cluster

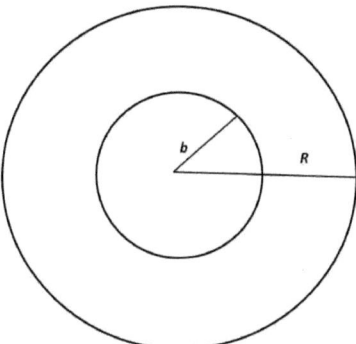

Figure 3.1: The outer sphere at $r = R$ is reflective and the inner sphere is absorbing at $r = b$.

and the Laplace transform of $\frac{dP(r,t)}{dr}$ is

$$\mathcal{L}\left(\frac{\partial P(r,t)}{\partial r}\right) = s\mathcal{L}(P(r,t)) - P(r,0). \tag{3.3}$$

To simplify the result we define a dimensionless radial coordinate $x = r\sqrt{s/D}$, to rewrite the diffusion in the form [58]:

$$P''(x,s) + \frac{d-1}{x}P'(x,s) - P(x,s) = -\frac{s^{(d-2)/2}}{D^{d/2}}\frac{\delta(x-x_0)}{\Omega_d x_0^{d-1}}, \tag{3.4}$$

The prime means differentiation with respect to x.

For each subdomain $x > x_0$ and $x < x_0$, this is a Bessel equation in which the solution is a superposition of the combinations $x^\nu I_\nu(x)$ and $x^\nu K_\nu(x)$ with $\nu = 1-d/2$. Where $I_\nu(x)$ and $K_\nu(x)$ are modified Bessel functions of the first and the second kind. First we apply the boundary conditions by considering absorbing and reflecting boundary conditions in one dimension to restrict the form of the Green's function. Boundaries are absorbing at $r = b$ and reflecting at $r = R$. Therefore we have $P(b,s) = 0$ and $\frac{dP(r,s)}{dr}|_{r=R} = 0$. Also the propagator should be continuous at $r = r_0$ or in our case $x = x_0$. In both regions, interior ($x_- \leq x < x_0$) and exterior ($x_0 \leq < x < x_+$), Green's function vanishes at their respective boundaries [58]. If $x_b = b\sqrt{s/D}$, then we have,

$$\begin{aligned}
P_<(x,s) &= Ax^\nu I_\nu(x) + Bx^\nu K_\nu(x), \\
P(x_b,s) &= 0, \\
P(x_b,s) &= b^\nu(AI_\nu(bx) + BK_\nu(b)) = 0, \\
B &= -A\frac{I_\nu(x_b)}{K_\nu(x_b)}
\end{aligned}$$

43

Then $P_<(x,s)$ is
$$P_<(x,s) = c_1 x^\nu C_<(x, x_b) \tag{3.5}$$
where $C_<(x, x_b) = I_\nu(x) K_\nu(x_b) - K_\nu(x) I_\nu(x_b)$. We apply reflecting boundary condition. If $x_R = R\sqrt{s/D}$, and we know $I'_\nu(x) = \frac{-\nu}{x} I_\nu(x) + I_{\nu-1}$ and $K'_\nu(x) = \frac{-\nu}{x} K_\nu(x) - K_{\nu-1}$ [93] we obtian

$$\begin{aligned} P_>(x,s) &= A x^\nu I_\nu(x) + B x^\nu K_\nu(x), \\ \frac{dP(x,s)}{dx}\Big|_{x=x_R} &= 0, \\ \frac{P(x,s)}{dx}\Big|_{x=x_R} &= A(x_R^\nu I_{\nu-1}(x_R)) - B(x_R^\nu K_{\nu-1}(x_R)), \quad B = \frac{I_{\nu-1}(x_R)}{K_{\nu-1}(x_R)}, \end{aligned} \tag{3.6}$$

Then $P_>(x,s)$ is
$$P_>(x,s) = c_2 x^\nu D_>(x, x_R). \tag{3.7}$$
where $D_>(x, x_R) = I_\nu(x) K_{\nu-1}(x_R) + K_\nu(x) I_{\nu-1}(x_R))$. The general solution for the Eq.(3.4) is:
$$P(x,s) = P_<(x,s) + \Theta(x - x_0)(P_>(x,s) - P_<(x,s)) \tag{3.8}$$
To ensure the continuity of the Green's function at $x = x_0$, $P_>(x,s)$ and $P_<(x,s)$ should be
$$P_>(x,s)|_{x=x_0} = P_<(x,s)|_{x=x_0}$$
therefore we obtain:
$$c_2 = c_1 \frac{C_<(x_0, x_b)}{D_>(x_0, x_R)}$$
To find c_1, we integrate Eq.(3.4) across the discontinuity at x_0 to give the joining condition
$$P'_>|_{x=x_0} - P'_<|_{x=x_0} = -\frac{-s^{(d-2)/2}}{D^{d/2}} \frac{1}{\Omega_d x_0^{d-1}}. \tag{3.9}$$
The derivative of $P_<(x,s)$ for $x = x_0$ is
$$P'_<|_{x=x_0} = \nu x_0^{\nu-1} c_1 C_<(x_0, x_b) + x_0^\nu c_1 C'_<(x_0, x_b), \tag{3.10}$$
and the derivative of $P_>(x,s)$ for $x = x_0$ is
$$P'_>|_{x=x_0} = \nu x_0^{\nu-1} c_2 D_>(x_0, x_R) + x_0^\nu c_2 D'_>(x_0, x_b), \tag{3.11}$$
From Eq.(3.9) we have
$$c_1 = \frac{1}{\Omega_d x_0^{d-1}} \frac{D_>(x_0, x_R)}{C_<(x_0, x_b) D'_>(x_0, x_R) - C'_<(x_0, x_b) D_>(x_0, x_R)}. \tag{3.12}$$
Therefore
$$P_<(x,s) = \frac{x_0^{-\nu} x^\nu}{\Omega_d x_0^{d-1}} \frac{C_<(x_0, x_b)}{C_<(x_0, x_b) D'_>(x_0, x_R) - C'_<(x_0, x_b) D_>(x_0, x_R)} C_<(x, x_b) \tag{3.13}$$

3.1 First passage properties between two spheres Diffusion in percolation cluster

and

$$P_>(x,s) = \frac{x_0^{-\nu} x^\nu}{\Omega_d x_0^{d-1}} \frac{D_>(x_0, x_R)}{C_<(x_0, x_b) D'_>(x_0, x_R) - C'_<(x_0, x_b) D_>(x_0, x_R)} D_>(x, x_R) \quad (3.14)$$

The total current for the absorbing inner sphere is the Laplace transform of the first passage probability, $J(x_0) = \int D \frac{\partial P}{\partial x}|_{x=x_b}$.

$$J(x,s) = D x^{d-1} \frac{\partial}{\partial x} P_<(x,s)|_{x=x_b} \left(x_b^{d-1} A_d\right) \quad (3.15)$$

It would be interesting to study the first passage characteristic in the long time limit when the outer sphere is enough large $x_R \to \infty$. In this limiting case, then we can say $K_\nu(x_R) \to 0$ so

$$J(x,s) \to \left(\frac{x_0}{x_b}\right)^\nu \frac{K_\nu(x_0)}{K_\nu(x_b)} \quad (3.16)$$

From the long time or small s behavior of this function we obtain the time dependent asymptotic. We study the basic case in details when

1. $\nu > 0$ and $d < 2$. The expansion of Eq.(3.16) gives

$$J(x) = 1 - \left(\frac{s}{4D}\right)^\nu \frac{\Gamma(1-\nu)}{\Gamma(1+\nu)} (x_b^{2\nu} - x_0^{2\nu}) - \left(\frac{s}{4D}\right) \frac{\Gamma(1-\nu)}{\Gamma(2-\nu)} (x_b^2 - x_0^2). \quad (3.17)$$

The probability of eventually reaching the sphere is 1, because the coefficient of zeroth power of s is one. Also the first passage time has a long tail of $t^{-(1+\nu)}$ that gives the infinite first passage time to the sphere.

2. $\nu = 0$ and $d = 2$. The expansion of Eq.(3.16) is

$$J(x) = 1 - 2\frac{\ln(x_0/x_b)}{\ln s} + \ldots \quad (3.18)$$

The probability of eventually hitting the circle is 1. The first correction varying as $1/\ln s$ corresponds to the first-passage probability having $1/(t \ln t^2)$ time dependence.

3. $\nu < 0$, and $d > 2$. We define $\nu' = -\nu > 0$, and the expansion of Eq.(3.16) gives

$$J(x) = \left(\frac{x_b}{x_0}\right)^{2\nu'} \left[1 - \left(\frac{s}{4D}\right)^{\nu'} \frac{\Gamma(1-\nu')}{\Gamma(1+\nu')} (x_b^{2\nu'} - x_0^{2\nu'}) - \left(\frac{s}{4D}\right) \frac{\Gamma(1-\nu')}{\Gamma(2-\nu')} (x_b^2 - x_0^2) \right] \quad (3.19)$$

From the $s \to 0$ limit of this expansion, we see the probability of eventually hitting the sphere is $(x_b/x_0)^{d-2}$. The leading correction term varying as $s^{\nu'}$ means that the hitting probability at time t varies as $t^{1+\nu'}$.

3.2 Time dependent first passage properties on fractals

Now we expand the problem to the diffusion on the fractals. The goal is to determine the full time dependency of the first passage probability between two concentric spheres on a random fractal. The first arrival probability to a point r for our particle in this medium becomes

$$J(r,t) = K r^{d_f - d_w + 1} \frac{\partial}{\partial r} p(r,t). \tag{3.20}$$

d_f is the fractal dimension, d_w denotes the random walk dimension, and $p(r,t)$ is the propagator. The radius dependent diffusion constant is $K(r) = K r^{-\theta}$, where $K(r) r^{d_f - 1}$ is defined as the total conductivity of a shell of $r^{d_f - 1}$ sites. The diffusion equation is

$$\frac{\partial}{\partial t} p(r,t) = \frac{1}{r^{d_f - 1}} \frac{\partial}{\partial r} \left(K(r) r^{d_f - 1} \frac{\partial}{\partial r} p(r,t) \right). \tag{3.21}$$

When d_f has an integer value $K(r) = K$ where K is the constant diffusion coefficient. According to [94] as an approximation, $K(r)$ can be written as $K(r) \sim K r^{-\theta}$ with $\theta = d_w - 2$ is identified as the anomalous diffusion exponent. The exact solution for diffusion on fractals is presented in [95]. We solve this equation exactly as we solved Eq.(3.1), by Laplace transformation and using the initial condition for $p(r,0) = \delta(r - r_0)/A_d r_0^{d_f - 1}$ and the boundary conditions (P(b,s)=0, $\frac{\partial}{\partial r} p(r,s) |_{r=R} = 0$), the solution becomes:

$$P(r,s) = P_<(r,s) + \Theta(r - r_0)(P_>(r,s) - P_<(r,s)). \tag{3.22}$$

Here $P_<$ is the solution for the region $b < r < r_0$, and $P_>$ the solution for $r_0 < r < R$,

$$P_<(r,s) = -\frac{D_>(r_0, R)}{K A_d r_0^{\alpha + d_f - d_w + 1}(C_<(r_0, b) D'_>(r_0, R) - C'_<(r_0, b) D_>(r_0, R))} r^\alpha C_<(r, b), \tag{3.23}$$

and

$$P_>(r,s) = -\frac{C_<(r_0, b)}{K A_d r_0^{\alpha + d_f - d_w + 1}(C_<(r_0, b) D'_>(r_0, R) - C'_<(r_0, b) D_>(r_0, R))} r^\alpha D_>(r, R). \tag{3.24}$$

We used the abbreviations

$$C_<(r,b) = I_\nu(\mu r^\gamma) K_\nu(\mu b^\gamma) - K_\nu(\mu r^\gamma) I_\nu(\mu b^\gamma) \tag{3.25}$$

$$D_>(r,R) = I_\nu(\mu r^\gamma)((\gamma \nu - \alpha) K_\nu(\mu R^\gamma) + \mu \gamma R^\gamma K_{\nu-1}(\mu R^\gamma))$$
$$+ K_\nu(\mu r^\gamma)((\alpha - \gamma \nu) I_\nu(\mu R^\gamma) + \mu \gamma R^\gamma I_{\nu-1}(\mu R^\gamma)), \tag{3.26}$$

and the parameters

$$\alpha = \frac{d_w - d_f}{2}, \nu = 1 - \frac{d_f}{d_w}, \mu = \frac{2\sqrt{s}}{d_w\sqrt{K}}, \gamma = \frac{d_w}{2}. \tag{3.27}$$

The propagator is zero at the surface of the absorbing sphere and its derivative is zero at the surface of the reflecting sphere.

The current density on the inner (absorbing) sphere (for simplicity we assume that the outer sphere is very large, $R \to \infty$) is

$$J(s)|_{R \to \infty} = \left(\frac{r_0}{b}\right)^{\nu\gamma} \frac{K_\nu(\mu r_0^\gamma)}{K_\nu(\mu b^\gamma)}. \tag{3.28}$$

To study the behavior of J at long or short times, we expand it for small or large values of s. For small values of s we find:

1. For $\nu = 0$

$$J(s)|_{R \to \infty} = \left[1 - 2\gamma \frac{\ln(\frac{r_0}{b})}{\ln(\frac{K}{s})}\right]. \tag{3.29}$$

   Similar to the results for normal space (Eq.(3.17)), the probability of hitting the inner sphere equals 1, and the first correction varying as $1/\ln s$ corresponds to the first passage probability having an asymptotic $1/(t \ln^2 t)$ time dependence.

2. For $\nu > 0$

$$J(us)|_{R \to \infty} = 1 + \left(\frac{s}{4K\gamma^2}\right)^\nu \frac{\Gamma(1-\nu)}{\Gamma(1+\nu)} \left(b^{2\gamma\nu} - r_0^{2\gamma\nu}\right)$$
$$- \left(\frac{s}{4K\gamma^2}\right) \frac{\Gamma(1-\nu)}{\Gamma(2-\nu)} \left(b^{2\gamma} - r_0^{2\gamma}\right) + ... \tag{3.30}$$

   From the leading correction term we see that the first passage density has the long time tail $t^{-\nu-1}$, leading to an infinite first passage time to the sphere.

3. For $\nu < 0$, $\beta = -\nu$

$$J(s)|_{R \to \infty} = \left(\frac{b}{r_0}\right)^{2\gamma\beta} \left(1 + \left(\frac{s}{4K\gamma^2}\right)^\beta \frac{\Gamma(1-\beta)}{\Gamma(1+\beta)} \left(b^{2\gamma\beta} - r_0^{2\gamma\beta}\right)\right.$$
$$\left. - \left(\frac{s}{4K\gamma^2}\right) \frac{\Gamma(1-\beta)}{\Gamma(2-\beta)} \left(b^{2\gamma} - r_0^{2\gamma}\right) + ...\right). \tag{3.31}$$

   If $s \to 0$, and if we assume that we are in two dimensional space ($d_f = 2$), when $\nu < 0$ the probability of hitting the sphere is $(b/r_0)^{d-2}$, as expected by

3.2 Time dependent FPT on fractals — Diffusion in percolation cluster

electrostatics. The correction terms give the time dependence of the hitting probability for the subset of particles that eventually reach the inner sphere. The leading correction term varying as s^β means that the hitting probability varies as $t^{-\beta-1}$, that is a power law of slope $1-\nu$ as shown in Fig. 3.2. Finally for short times, for $s \to \infty$

$$J(s)|_{R\to\infty} = \left(\frac{r_0}{b}\right)^{\gamma(\nu-\frac{1}{2})} e^{-\frac{\sqrt{s}}{\gamma\sqrt{K}}(r_0^\gamma - b^\gamma)} \left(1 - \frac{\gamma\sqrt{K}\,(4\nu^2-1)}{8\sqrt{s}}\left(b^{-\gamma} - r_0^{-\gamma}\right) + ...\right). \quad (3.32)$$

To plot the current $J(t)$ we need to calculate the inverse Laplace transform of Eq.(3.32). We used a Mathematica package *NumericalLaplaceInversion.m* [96] to do the inverse laplace transform numerically. In Fig.3.2 the current density $J(t)$ versus time when $R \to \infty$ in a fractal space with fractal dimension of $d_f = 3.5$ and $\nu = 0$ is plotted.

In 3 dimensional percolation cluster $d_w > d_f$ and the walk is compact and leads to

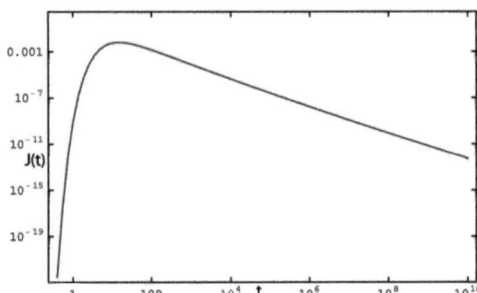

Figure 3.2: The Current density versus time, when $R \to \infty$, $d_f = 3.5, \theta = 1.5$, $\nu = 0, r_0 = 6$.

$r_0^{d_w - d_f}$ behavior of the MFPT [97]. That r_0 is the distance between the target and the starting point for the random walker. Our results in Eq.(3.31) are in agreement with [97] and the MFPT is proportional to $r_0^{-2\gamma\beta} = r_0^{d_w - d_f}$. This feature has strong implications on reaction kinetics in cells. In the cases where the fractal description of the cell environment rules, our results show that reaction times crucially depend on the source target distance r_0. On the example of gene colocalization, the importance of dependence on the starting point has been shown [98]. On the other hand, in processes in which the CTRW description of transport is valid, MFPTs do not depend on the starting point at large distance r_0 [97].

3.3 Simulation of Random walk

The calculation of the first passage time for the full problem is not trivial. Apart from the results for the characteristic time, further quantities of interest at finite cell radius R can only be obtained numerically. We simulate the diffusion on a critical percolation cluster to model the subdiffusion caused by macromolecular crowding in cytoplasm. In the percolation cluster, the bonds will connect points of a lattice with the probability greater than a critical probability (p_c). In two dimension $p_c = 0.5$ and in three dimension $p_c = 0.2488$, for hypercubic lattices.

A random walk is then performed on the biggest cluster of the lattice to make sure that the cluster connects the sides of the lattice, see the example in Fig. 3.3b. When p reaches p_c there will be at least one cluster that is connected the sides of the lattice together. We use labeling method [99] to count the number of sites in each cluster and finally to find the largest cluster and delete the bonds that do not belong to the largest cluster. In appendix A a part of the source code that describes the generating percolation cluster and finding the infinite cluster is attached.

Random walker chooses one of $2d$ possible directions to move. The move to the new point of the lattice can be done by choosing a random number between 0 and 1. There are two methods used to generate random numbers. One is physical methods that measures some random physical phenomenon and then compensates for possible biases in the measurement process. The other method is often called pseudorandom number generators, uses computational algorithms that produce long sequences of random results, which are determined by a shorter initial value, known as a seed. The term "random number generator" usually refers to uniform random number generator, while other distributions like Gaussian distribution are also possible. For our case, it is essential to generate random number with uniform distribution to ensure the equal probability for jumps. In our code we used uniform random number generator (ran3) to generate the random numbers. In Fig. 3.3a, the random walker moves to its right if the generated random number is between zero and $\frac{1}{2d}$. In order to have different random walk paths we need different random number sequences, therefore it is essential to change the seed every time we run the code. The seed can be a function of time then it changes automatically. For example the function secnds(x) in Fortran, gets the time in seconds from the real-time system clock. X is a reference time, in seconds. In the function ran3, seed should be large negative integer, therefore we set it as -2secnds(x)+1.

In the regular lattice all the bonds are available for the random walker and the motion normal Brownian. The random walk on the percolation cluster is similar to blind ant walk [100] as the random walker chooses a direction as described above and then checks if the chosen bond is available. If the chosen bond is available the random walker moves to the new point of the lattice and we count one time step, if the chosen bond is not available the random walker stays at its place and we count

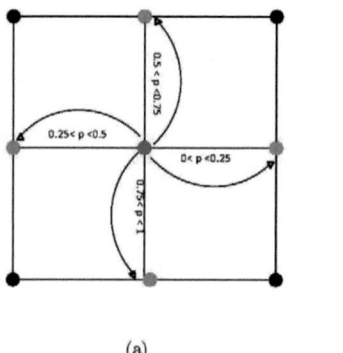 

(a) (b)

Figure 3.3: (a)A Random walk on two dimensional regular lattice. The random walker is in the center of the lattice, and with probability p, that is a generated random number, moves to its on of four nearest neighboring sites. If $0 < p < 1/4$ it moves to to right, if $1/4 < p < 1/2$ it moves to left, if $1/2 < p < 3/4$, it moves up, and if $3/4 < p < 1$ it moves down. (b) Random walk path on the infinite cluster in the lattice of size 35 by 35, after 10000 time steps.

one time step. In Fig. 3.3a, a random walk on two dimensional regular lattice is shown. The random walker starts in the middle of the lattice and has four choices with equal probabilities to move. In Fig. 3.3b random walk path on the infinite cluster in the lattice of size 35 by 35, after 10000 time steps is displayed. Although the random walker has taken sufficient number of steps, it has not visited all the site of the square lattice.

3.3.1 Analysis of the mean squared displacement

Let us first study the random walk by studying the mean squared displacement and first passage time density on regular lattices and percolation clusters. As we had discussed it before the diffusion on percolation cluster is similar to diffusion on fractals and it is an anomalous diffusion. In this simulation we calculate the mean squared displacement $\langle r^2 \rangle \sim t^{2/d_w}$ to check if the motion is subdiffusive and we obtain the mean first passage time to the target. When the motion of random walker is a normal diffusive motion, the MSD should be linear versus time. We plotted ensemble averaged and time averaged MSD versus time. Both graphs are linear in time and in finite system, both reach a plateau when the random walker visits all the points of the lattice.

In order to calculate the expectation value of x^2 for a particle diffusion between two reflecting boundaries at $x = \pm a$ in one dimension, we should find the propagator. If the particle starts at $x_0 = 0$, the diffusion equation is;

$$\frac{\partial P(x,t)}{\partial t} = D\frac{\partial^2 P(x,t)}{\partial x^2}, \tag{3.33}$$

and we have two boundary conditions, the reflection boundary condition at $x = \pm a$

$$\frac{\partial P(\pm a, t)}{\partial x} = 0, \tag{3.34}$$

and for initial condition

$$P(x,0) = \delta(x), \tag{3.35}$$

then we can separate the variables and define the propagator as

$$P(x,t) = T(t)X(x), \tag{3.36}$$

that $X(x)$ is

$$X(x) = \sum_{i=0}^{n} A_n \sin(k_n x) + B_n \cos(k_n x). \tag{3.37}$$

If we substitute Eq.(3.36) to Eq.(3.33), we will have;

$$T(t) = e^{C_n t}. \tag{3.38}$$

We obtain C_n

$$C_n = -Dk_n^2. \tag{3.39}$$

by applying the reflecting boundary condition, we have

$$\frac{\partial P(a,t)}{\partial x} = 0, \quad \sum_{i=0}^{n} B_n \cos(k_n x) = 0, \quad k_n = n\pi/a, \tag{3.40}$$

and

$$\frac{\partial P(-a,t)}{\partial x} = 0, \quad \sum_{n=0}^{\infty} A_n \cos(k_n x) = 0, \quad A_n = 0. \tag{3.41}$$

Then $P(x,t) = T(t)X(x)$

$$P(x,t) = \sum_{i=0}^{n} B_n \cos\left(\frac{n\pi x}{a}\right) e^{-D(n\pi/a)^2 t}. \tag{3.42}$$

To apply the initial conditions, we should expand the delta function. We know the fourier expansion of a function in an interval [-a,a] is $f(x) = B_0/2 + \sum B_n \cos(\frac{n\pi x}{a}) +$

51

3.3 Simulation of Random walk Diffusion in percolation cluster

$\sum A_n \sin(\frac{n\pi x}{a})$. That $B_0 = \frac{1}{a}\int_{-a}^{a} f(x)dx$ and $B_n = \frac{1}{a}\int_{-a}^{a} f(x)\cos(\frac{n\pi x}{a})dx$. Therefore for delta function, we have

$$B_0 = \frac{1}{a}\int_{-a}^{+a} \delta(x)dx = \frac{1}{a}, \quad (3.43)$$

and

$$B_n = \frac{1}{a}\int_{-a}^{+a} \delta(x)\cos(\frac{n\pi x}{a})dx = \frac{1}{a}. \quad (3.44)$$

As A_n is zero, we have

$$\delta(x) = \frac{1}{2a} + \frac{1}{a}\sum_{n=1}^{\infty} \cos\left(\frac{n\pi x}{a}\right). \quad (3.45)$$

Then Eq.(3.46) will be

$$P(x,t) = \frac{1}{2a} + \frac{1}{a}\sum_{n=1}^{\infty} \cos\left(\frac{n\pi x}{a}\right) e^{-D(n\pi/a)^2 t}. \quad (3.46)$$

The expectation value of x^2 is:

$$\langle x^2 \rangle = \int_{-a}^{+a} \frac{1}{2a}x^2 dx + \int_{-a}^{+a} x^2 \frac{1}{a}\sum_{n=1}^{\infty} \cos\left(\frac{n\pi x}{a}\right) e^{-D(n\pi/a)^2 t} dx. \quad (3.47)$$

In long times, the second part of above equation will be zero. And the first part is:

$$\int_{-a}^{+a} \frac{1}{2a}x^2 dx = \frac{a^2}{3}. \quad (3.48)$$

Therefore we should observe a plateau in MSD after a long time that reaches the values $\frac{a^2}{3}$ in one dimension. In case of two dimensional random walk in a square lattice of size a, the plateau reaches the value $\frac{2a^2}{3}$ and the plateau for three dimensional walk reaches the values a^2. For shorter time we define the time first derivative of the second part of the Eq.(3.47) as a Jacobi theta function $\theta_4(z,q)$, when $z = 0$.

$$\frac{d}{dt}\frac{1}{a}\sum_{n=1}^{\infty} \frac{4a^3}{n^2\pi^2} \cos(n\pi) e^{-D(n\pi/a)^2 t} = -4D\sum_{n=1}^{\infty} \cos(n\pi) e^{-D(n\pi/a)^2 t} \quad (3.49)$$

The Jacobi theta function $\theta_4(z,q)$ is defined as

$$\theta_4(z,q) = 1 + \sum_{n=1}^{\infty} (-1)^n q^{n^2} \cos(2\pi z) \quad (3.50)$$

3.3 Simulation of Random walk — Diffusion in percolation cluster

Therefore Eq.(3.49) is $-4D\frac{\theta_4(0,q)-1}{2}$ with $q = e^{-D\pi^2 t/a^2}$. For short times, if $t \to 0$, then $q \to 1$ and $\theta_4(1,0) = 0$. Therefore Eq.(3.49) will be $2D$. In short times when $t \to 0$ we know $\lim_{n \to 0}\langle x^2 \rangle = 0$, therefore the second part of Eq.(3.47) at $t = 0$ is $\frac{1}{a}\sum_{n=1}^{\infty} \frac{4a^3}{n^2\pi^2} \cos(n\pi) = \frac{1}{a}\frac{4a^3}{\pi^2}\left(-\frac{\pi^2}{12}\right) = -\frac{a^2}{3}$. Finally the mean squared displacement in one dimension is

$$\langle x^2 \rangle = 2Dt. \qquad (3.51)$$

In Fig. 3.4a the mean squared displacement (time averaged mean squared displacement) for a random walk on a two dimensional regular lattice is plotted. The size of the lattice is 20 by 20 ($a = 10$ in Eq.(3.47)). The slope of MSD is proportional to t, for short times and after 100 steps it reaches the plateau that according to Eq.(3.47) it should have the value of $2a^2/3 = 200/3$. Fig. 3.4b is the MSD for the same three dimensional lattice.

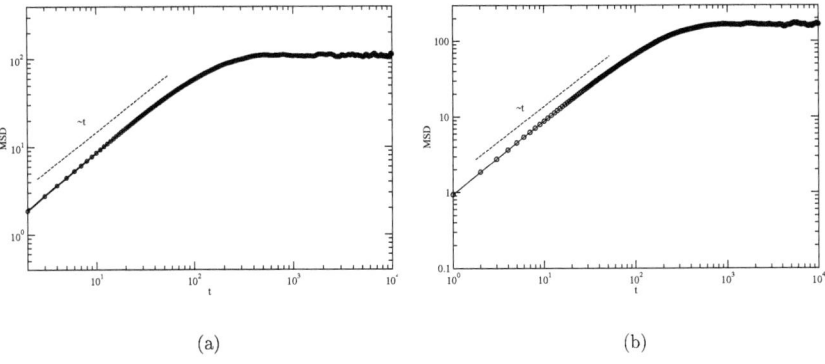

(a) (b)

Figure 3.4: (a) The mean squared displacement, on a two dimensional regular lattice (20 by 20)versus time steps. The slope of the MSD is proportional to t for short times and reaches the plateau $2a^2/3 = 200/3$. (b) The mean squared displacement, on a three dimensional regular lattice (20 by 20 by 20)versus time steps. The slope of the MSD is proportional to t that proves the normal diffusion on the regular lattice.

In Fig. 3.5 the mean squared displacement for a random walk on percolation cluster in two and three dimensions is plotted. The diffusion is not normal anymore and is anomalous. The Mean squared displacement is always proportional to t^{2/d_w} in the case of normal diffusion d_w is equal to 2, therefore MSD will be linear to time.

3.3 Simulation of Random walk — Diffusion in percolation cluster

In case of subdiffusion on the two dimensional percolation cluster from chapter one $d_w = 2.87$ then the MSD is proportional to $t^{0.69}$ Fig. 3.5a. The walk dimension on three dimensional percolation cluster is $d_w = 3.88$ therefore MSD is proportional to $t^{0.515}$ (Fig. 3.5b). Both mean squared displacements reach a plateau of $(2a^2/3)$ in two dimensional random walk (Fig. 3.5a and a^2 in three dimensional random walk on percolation cluster (Fig. 3.5b).

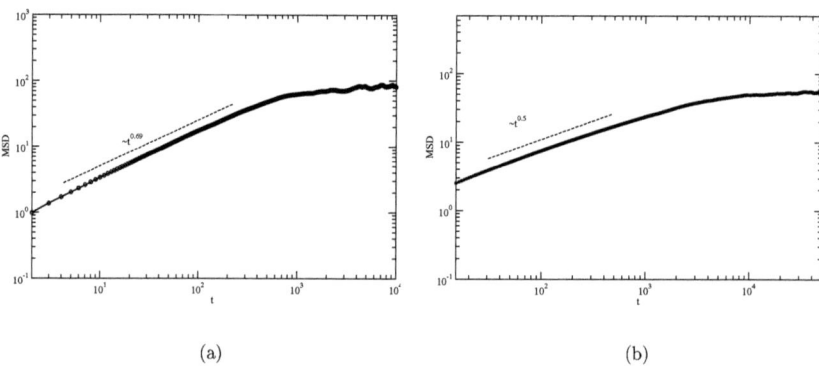

(a)   (b)

Figure 3.5: (a) The mean squared displacement for a random walk on bond percolation cluster in two dimension. The diffusion is anomalous. The Mean squared displacement is proportional to t^{2/d_w} and in this case is proportional to $t^{0.69}$. (b) The mean squared displacement for a random walk on percolation cluster in three dimensions. MSD is proportional to $t^{0.51}$.

3.3.2 Analyzing the first passage time density

The First passage time distribution can also give a clue to the diffusion type [58]. Usually we place a target (absorbing boundary condition) and as soon as the random walker hits the target we record the first passage time. After several iterations we can plot the first passage time distribution (FPTD). In an infinite lattice the slope of FPTD is a power law and proportional to $t^{-1-\alpha}$, that $\alpha = 2/d_w$ is the anomalous diffusion exponent. Normally to plot a FPT distribution we should count the number of observations n in an interval a. If a is a constant interval we will have too many fluctuations at the tail (right hand side) of the curve. At the right hand side of the curve, each bin (the interval that the number of event are counted in) only has very few samples in, so the fluctuations in the bin counts are large and this appears as a noisy curve on the plot. To have a FPTD with out too many fluctuation, we can

3.3 Simulation of Random walk   Diffusion in percolation cluster

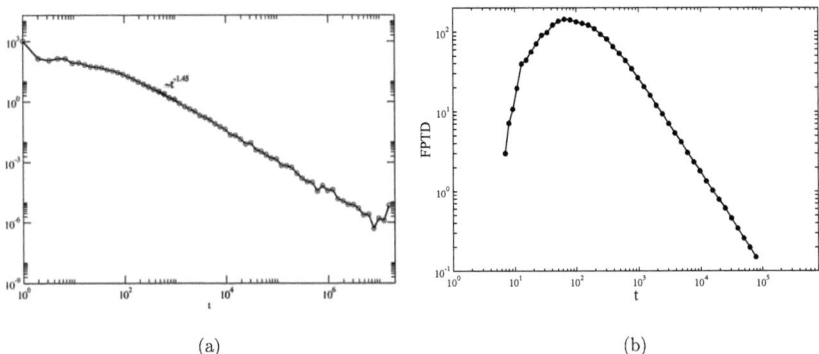

(a)   (b)

Figure 3.6: (a)The First passage time density of a random walk in one dimension, the target is an absorbing point is one dimensional lattice, FPTD should be power law with the slope of $t^{-3/2}$. (b) First passage time density of a random walk in two dimensional regular lattice, the target is an absorbing square of size 2 by 2 lattice, FPTD should be logarithmic in two dimension (see Eq.(3.18)).

change the binning to the log bin and normalize it by dividing the FPTD to size of each bin. Otherwise the slope of the curve will be the real slope plus one. By definition a bin of constant logarithmic width means that the logarithm of the upper edge of a bin (t_{i+1}) is equal to the logarithm of the lower edge of that bin (t_i) plus the bin width (b). That is,

$$\log(t_{i+1}) = \log(t_i) + b,$$

$$t_{i+1} = e^{\log(t_i)+b} = t_i e^b, \tag{3.52}$$

the linear bin width of bin i, w_i, is defined as $w_i = t_{i+1} - t_i$. The linear bin width is directly proportional to t_i because

$$w_i = t_i e^b - t_i = t_i(e^b - 1). \tag{3.53}$$

The number of observations in a bin (n) is equal to the density of observations in that bin times the width of that bin. Therefore if the probability density function be, $f(t) \propto t^\lambda$ and the width of the bin, then $w \propto t$

$$n \propto t^\lambda t = t^{\lambda+1}, \tag{3.54}$$

and regressing $\log(n)t^{\lambda+1}$ versus $\log(t)$ yields a slope equal to $\lambda + 1$, not λ. If n is divided by the linear width of the bin then,

$$\frac{n}{w} = \frac{t^{\lambda+1}}{t} = t^\lambda, \tag{3.55}$$

3.3 Simulation of Random walk   Diffusion in percolation cluster

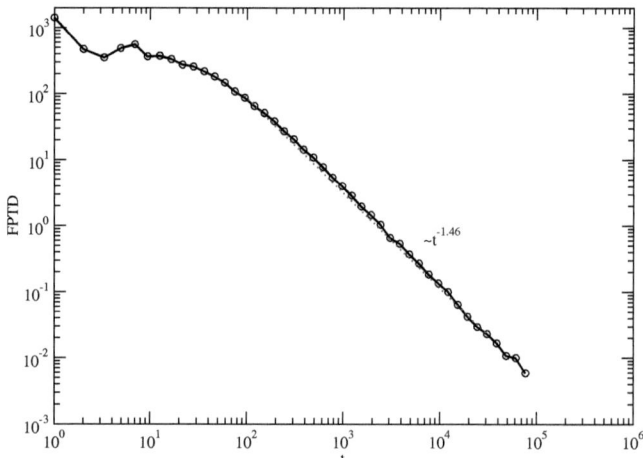

Figure 3.7: The first passage time density of a random walk in three dimensional regular lattice. The target is an absorbing cube of size 2, in the center of a very big lattice. FPTD is a power law with the slope of $t^{-3/2}$

and therefore a regression of the normalized logarithmic bin counts against the logarithm of t will estimate λ.

The other solution to omit the fluctuations is plotting the cumulative distribution function. The cumulative probability of a power law probability distribution is also power law but with an exponent $\lambda - 1$. The cumulative distribution function of $f(t)$, describes the probability that a real-valued random variable T with a given probability distribution will be found at a value less than or equal to t, in our case cumulative function will be survival probability.

$$F(t) = \int_t^\infty f(t')dt', \qquad (3.56)$$

$f(t)$ is the FPTD and usually a power law Ct^λ.

$$F(t) = \int_t^{+\infty} Ct^\lambda dt' = \frac{C}{\lambda - 1} x^{-(\lambda - 1)}. \qquad (3.57)$$

56

3.3 Simulation of Random walk — Diffusion in percolation cluster

To have the cumulative (or survival probability), we assume the number of events (found targets in that time interval) in the first time interval, out of total number of events N is m_1, then the cumulative distribution for that interval can be calculated easily as the probability that the target is not found yet $(N - m_1)$. Therefore if the number of events in the second time interval is m_2, the survival probability will be $N - m_1 - m_2$. Later we present some examples that calculating the cumulative distribution provides a more exact results.

Fig. 3.6a shows the first passage time density of a random walk in one dimension, the target is an absorbing point in one dimensional lattice, and after the random walker hits the target the search is terminated and the search time will be recorded. FPTD should be power law with the slope of $t^{-3/2}$. Fig. 3.6b displays the first passage time density of a random walk in two dimensional regular lattice, the target is an absorbing square in the center of the lattice. Here we see that the slope of FPTD logarithmic according to Eq.(3.18). In Fig. 3.7 the first passage time density of a random walk in three dimensional regular lattice is displayed. The target is an absorbing cube of size 2 in the center of a large enough lattice and we see that FPTD is a power law with the slope of $t^{-3/2}$.

All the FPTDs are plotted using log bin. An example of FPTD with normal binnig is shown in Fig. 3.8. The fluctuations on the right hand side of the plot are too many and it makes it quite impossible to determine the precise value of he slope.

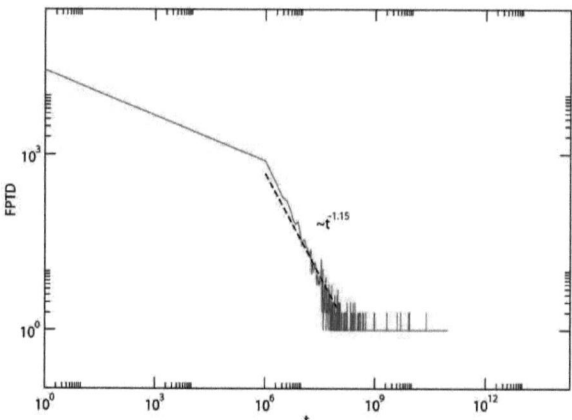

Figure 3.8: The first passage time density of a random walk using normal bin. There are too many fluctuations on the tail of the power law and the estimation of the slope is very difficult.

Chapter 4

Modeling EcoRV's dynamic in E.coli cell

Here we present a further clue to understanding the relation between crowding-induced anomalous diffusion and the design of vital cellular mechanisms. Our case study addresses the dynamics of the type II restriction endonuclease EcoRV, that occurs in the bacterium E.coli.

E.coli cells are typically rod-shaped, and are about 2 micrometers (μm) long and 0.5 μm in diameter, with a cell volume of $0.6 - 0.7$ (μm^3)[24] and E.coli DNA is mainly concentrated in the middle of the cell (Fig. 4.1). We mentioned in introduction that EcoRV enzyme recognizes the 6-base DNA sequence 5'-GAT|ATC-3' and makes a cut at the vertical line therefore renders it inactive with respect to transcription and replication. EcoRV forms a homodimer in solution before binding and acting on its recognition sequence [25]. Its molecular weight in solution is 58 kD [101], and thus belongs to the range of sizes for which subdiffusion under crowding was reported [15, 20]. Initially the enzyme binds weakly to a non-specific site on the DNA (nonspecific binding) and randomly walks along the molecule until the specific recognition site is found [26]. Then it binds to the specific site (specific binding) and cleavage occurs within the recognition sequence, and does not require ATP hydrolysis [26]. Fig. 4.2 shows how EcoRV cleaves the DNA. DNA cleavage is an important mechanism in the cellular defence against foreign DNA of viruses attacking the cell. The cell's own DNA is protected against EcoRV action by methylation by a modification enzyme of the DNA at cytosine or adenine [26]. Bacteria uses methylase to be able to differentiate between foreign genetic material and their own, therefore protecting their DNA from their own immune system.

Interestingly, as seen by X-ray crystallography, EcoRV can be found in two configurational states [27, 28]. The unbound protein may switch between an inactive structure with a closed cleft and another, in which the cleft is more open. In open state EcoRV non-specifically binds to DNA that could be the native DNA or foreign DNA. One should keep in mind that EcoRV can bind to the native DNA non

Modeling EcoRV's dynamic in E.coli cell

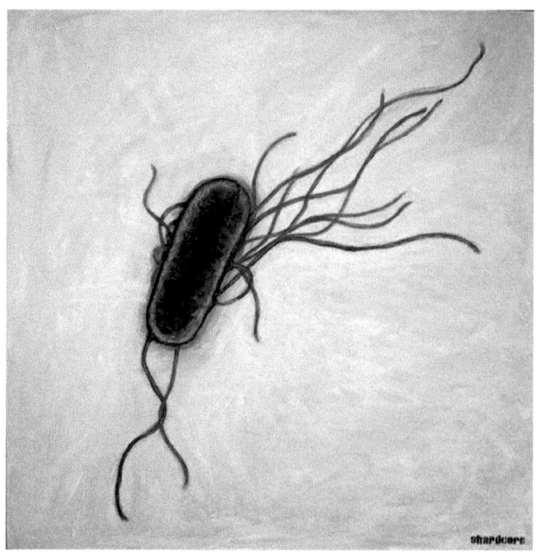

Figure 4.1: A painting from E.coli [102] E.coli lives in the human large intestine and assists with waste processing, vitamin K production and food absorbtion. The average human defecates between 100 billion and 10 trillion E.coli bacteria every day.

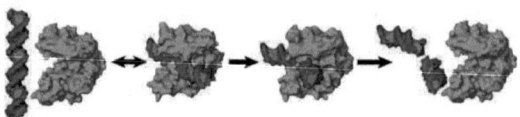

Figure 4.2: EcoRV cleaving DNA. The protein loosely binds DNA and scans for its recognition sequence. Once found, EcoRV kinks the DNA in a 50°angle and cleaves at the cognate sequence [103].

specifically but would not be able to cleave it due to methylation, and it can bind specifically and also non specifically to the foreign DNA. In Fig. 4.3 we sketch an *E.coli* cell and its native DNA, EcoRV enzymes in active and inactive states being either attached non-specifically to the native DNA or freely diffusing in the cellular cytoplasm. The apparent void space in reality is a highly crowded ('superdense' [16]) complex liquid, in which the enzymes subdiffuse. An invading DNA is being recognised and cleaved and deactivated by active EcoRV enzymes.

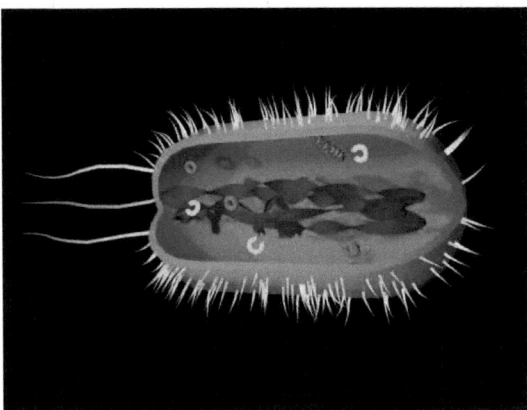

Figure 4.3: Sketch of an *E.coli* cell with native DNA (violet) concentrated in the center. EcoRV restriction enzymes occur in two isomers: inactive, with closed cleft (red), and active with open cleft (yellow). Invading, foreign DNA (red double-helix) is attacked by EcoRV and cut. The void intracellular space shown here in reality is crowded by larger biopolymers.

The probability to find the enzyme active x_{act} at a given instant of time is as low as $\sim 1\%$ [28, 29]. A large volume fraction of the cytoplasm is occupied by larger biopolymers [6](see Fig. 2). For example, the cytosol of E.coli contains about 300-400 milligrammes per millilitre (mg/ml) of macromolecules [11]. EcoRV in solution forms homodimers of molecular weight 58 kD [101], and thus belongs to the range of sizes for which subdiffusion under crowding was reported [15, 20].
We use different processes to model the EcoRV's dynamic in E.coli cell. The structure of the crowded bacterial cytoplasm resembles a random fractal [32] and we model the enzyme dynamics as a random walk on a discrete cubic lattice whose lattice constant a corresponds to the EcoRV size. To model the cytoplasmic crowding we use static bond percolation cluster. A number of recent works applied the percolation idea to stochastic motion in a crowded environment [36, 37]. Later we study the same system using different process to model the crowding inside the cell by using CTRW and fBm.

To obtain a better physical understanding of EcoRV non-specific binding to the cell's native DNA, we study the dependence of the MFPT on the non-specific binding constant K_{ns}^0. Experimentally, K_{ns}^0 can be varied by changing the salt concentration of the solution.

4.1 Theory

For the cubic lattice ($p = 1$) we obtain an analytical expression for the MFPT for the geometry sketched in Fig. 4.4. We distinguish Region 1 containing the native DNA, and Region 2 representing the cytoplasm. In this region foreign DNA enters and the EcoRV action occurs. Let us first address the non-specific binding of EcoRV

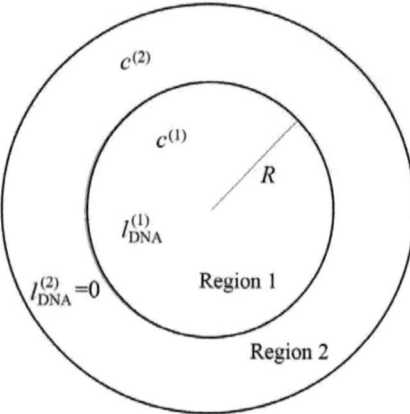

Figure 4.4: Sketch of the cross section of an E.coli model cell. Region 1 contains the cell's native DNA. In Region 2 ("cytoplasm"), foreign target DNA are attacked by active restriction enzymes. The various symbols are explained in the text.

enzymes to the native cellular DNA in Region 1, corresponding to a volume of length of L and radius R. Assuming rapid equilibrium with respect to enzyme binding and unbinding from the DNA, we observe the following relation between the volume concentrations of bound and unbound *active* (ready-to-bind) enzymes,

$$\frac{c_{\text{act}}^{(1)}}{c_{\text{bound}}^{(1)}} = \frac{1}{K_{\text{ns}}^0 l_{\text{DNA}}^{(1)}}. \tag{4.1}$$

In our notation $c^{(1)}$ denotes the overall volume concentration of enzymes in Region 1, while $c_{\text{bound}}^{(1)}$ and $c_{\text{bulk}}^{(1)}$, respectively, measure the volume concentrations of enzymes bound to the native DNA and of unbound enzymes. The non-specific binding constant K_{ns}^0 to DNA refers to active (open-cleft) enzymes per DNA length, and is of dimension $[K_{\text{ns}}^0] = [\text{M}^{-1}\text{bp}^{-1}]$. Finally, $l_{\text{DNA}}^{(1)}$ is the length of DNA per volume in Region 1. Of the unbound enzymes, a fraction x_{act} is in the active (open-cleft) state, ready to bind to DNA. Thus, the concentration of active enzymes in Region 1 becomes $c_{\text{act}}^{(1)} = x_{\text{act}} c_{\text{bulk}}^{(1)}$, and one may introduce an overall binding constant

$K_{\text{ns}} = x_{\text{act}} K_{\text{ns}}^0$ in Eq.(4.1):

$$\frac{c_{\text{bulk}}^{(1)}}{c_{\text{bound}}^{(1)}} = \frac{1}{x_{\text{act}} K_{\text{ns}}^0 l_{\text{DNA}}^{(1)}} = \frac{1}{K_{\text{ns}} l_{\text{DNA}}^{(1)}}. \tag{4.2}$$

As the total concentration of enzymes in Region 1 is $c^{(1)} = c_{\text{bulk}}^{(1)} + c_{\text{bound}}^{(1)}$, and we have $c_{\text{bound}}^{(1)} = c_{\text{bulk}}^{(1)} x_{\text{act}} K_{\text{ns}}^0 l_{\text{DNA}}^{(1)}$ we can write

$$c_{\text{bulk}}^{(1)} = \frac{c^{(1)}}{1 + x_{\text{act}} K_{\text{ns}}^0 l_{\text{DNA}}^{(1)}}. \tag{4.3}$$

In Region 1 the enzyme concentration will be governed by the diffusion equation of the form

$$\frac{\partial c^{(1)}}{\partial t} = D_{\text{eff}} \nabla^2 c^{(1)}, \tag{4.4}$$

where $D_{\text{eff}} = D_{\text{3d}}/(1 + x_{\text{act}} K_{\text{ns}}^0 l_{\text{DNA}}^{(1)})$ is an effective diffusion coefficient incorporating the assumption of rapid equilibrium with respect to binding to DNA and switching between active and dormant states. 1D diffusion along the DNA is assumed so slow that it can be ignored in connection with the overall diffusion of the enzyme. Indeed the 1D diffusion constant for EcoRV have been measured to be orders of magnitude smaller than for 3D diffusion [104].

In Region 2 (the "cytoplasm"), we assume that 3D diffusion is fast such that we can write a conservation law for enzymes in the form of the difference between what the flux across the boundary with Region 1 and the amount of enzymes reacting with the target per time,

$$\frac{d}{dt} V^{(2)} c^{(2)} = -A^{(1)} D_{\text{eff}} \left. \frac{\partial c^{(1)}}{\partial r} \right|_{r=R} - k_a c^{(2)}. \tag{4.5}$$

Here $V^{(2)}$ is the volume of Region 2, $A^{(1)} = 2\pi R L$ is the surface area of Region 1, and k_a is the rate constant for reaction with targets. The x_{act} dependence of k_a is $k_a = x_{\text{act}} k_a^0$, where k_a^0 is the rate constant for the active state. Note that in our approach we assume that the switching between active and dormant state is fast in comparison with the diffusion across the regions, i.e., we may assume an equilibrium between these two states. Finally, we take c_{bulk} to be continuous across the boundary between the Region 1 and Region 2, and that initially the system is at equilibrium with respect to the reaction-free situation with $k_a = 0$. From the above system of equations the average search time yields in the form

$$T = \left(1 + x_{\text{act}} K_{\text{ns}}^0 l_{\text{DNA}}^{(1)}\right) \left\{ \frac{V^{(1)}}{k_a^0 x_{\text{act}}}(1+y) + \frac{R^2}{8 D_{\text{3d}}} \frac{1}{1+y} \right\}, \tag{4.6}$$

where $y = V^{(2)} / \left[V^{(1)}(1 + x_{\text{act}} K_{\text{ns}}^0 l_{\text{DNA}}^{(1)})\right]$ and $V^{(1)}$ is the volume of Region 1. The simulations were carried out on a $100 \times 100 \times 100$ cubic lattice with native

DNA occupying a $100 \times 50 \times 50$ lattice in the middle. To compare the present calculation with the simulations we assume a lattice spacing of $a = 10\,\text{nm}$ and set $L = 1\,\mu\text{m}$, $V^{(2)} = 1\,\mu\text{m}^3$ and therefore $V^{(1)} = 0.25\,\mu\text{m}^3$. From this we obtain $R = \sqrt{V^{(1)}/(\pi L)} \approx 0.28\,\mu\text{m}$.
We choose the enzyme diffusivity $D_{3d} = 3\mu\text{m}^2/\text{sec}$. This value obtained for lac repressor in vivo at short times [105]. The DNA length per volume is $l_{\text{DNA}}^{(1)} = 1.5 \times 10^{-3}\text{m}/V^{(1)}$, and $K_{\text{ns}}^0 = 10^7 \text{M}^{-1}\text{bp}^{-1}$, such that $K_{\text{ns}} = 10^5 \text{M}^{-1}\text{bp}^{-1}$ when $x_{\text{act}} = 0.01$ [106] with the base pair length bp $= 0.35\,\text{nm}$.

For the target association rate constant we take $k_a^0 = (Na)^3/T_{\text{lattice}} = a^3(1 - R_{3d})/\tau_{\text{step}}$. Here $T_{\text{lattice}} = N^3 \tau_{\text{step}}/(1 - R_{3d})$ is the average search time for a random walker starting far from the target on a $N \times N \times N$ cubic lattice and spending a time τ_{step} per step to nearest neighbor sites. $R_{3d} \approx 0.340537$ is the walker's return probability to its origin [57].

To match the time step τ_{step} with the above diffusion constant through the mean squared displacement of the walker we obtain $\tau_{\text{step}} = a^2/(6 D_{3d}) \approx 5.6\,\mu\text{s}$. The assumption of a one lattice site target gives a target size of $a = 10\,\text{nm}$. We have chosen this target size to be lower than the in vitro effective sliding length [29] at optimal salt conditions, partly due to possible blocking on the DNA by other DNA-binding proteins.

The above numbers result in $K_{\text{ns}}^0 l_{\text{DNA}}^{(1)} \approx 3 \times 10^5$, $y|_{x_{\text{act}}=1} \approx 10^{-5}$, $y|_{x_{\text{act}}=0.01} \approx 10^{-3}$, $V^{(1)}/k_a^0 \approx 2\,\text{sec}$, and $R^2/(8 D_{3d}) \approx 0.003\,\text{sec}$, and with these parameters we have to a good approximation $T = K_{\text{ns}}^0 l_{\text{DNA}}^{(1)} V^{(1)}/k_a^0$, regardless of whether $x_{\text{act}} = 1$ or $x_{\text{act}} = 0.01$.

4.2 Percolation cluster application

Our simulations of the search process were carried out on a $100 \times 100 \times 100$ cubic lattice and generate a bond percolation on it. We considered bond percolation, i.e., for each pair of nearest neighbor sites a bond is constructed with a probability p. The largest cluster of connected sites was chosen and the remaining sites will be deleted. Of these remaining sites those sites that are within the central $100 \times 50 \times 50$-size part constitute Region 1 with the native DNA. The number of such sites is denoted by $N^{(1)}$ and the volume of this region is thus $V^{(1)} = a^3 N^{(1)}$. The remaining sites outside this region constitute the cytoplasm. The number of these sites is $N^{(2)}$, and the corresponding volume becomes $V^{(2)} = a^3 N^{(2)}$. A single target site is chosen randomly among the cytoplasm sites. The searching random walker is only allowed to walk between connected sites. In Fig. 4.5 a random walk on the bond percolation is illustrated. The initial position of the searching random walker is chosen by first

4.2 Percolation cluster application — Modeling EcoRV's dynamic in E.coli cell

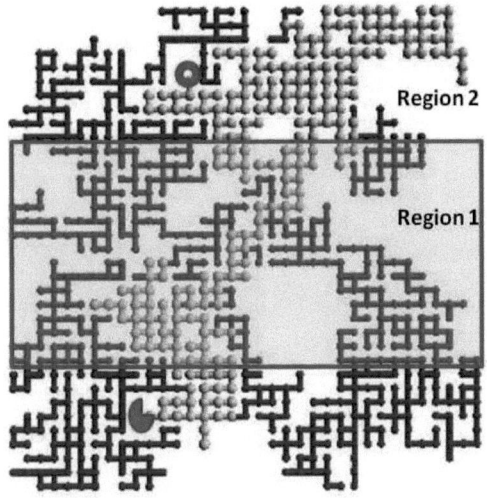

Figure 4.5: EcoRV's random walk (red dot), searching for the target (green dot), on percolation cluster modeling the crowding.

deciding whether it is bound or unbound. From Eqs.(4.1) and (4.2), we can say

$$\frac{c_{bound}^{(1)}}{c_{bulk}^{(1)}} = K_{ns} l_{DNA}^{(1)}) = x_{act} K_{ns}^0 l_{DNA}^{(1)}. \tag{4.7}$$

Also
$$V^{(1)} c_{bound}^{(1)} + V^{(1)} c_{bulk}^{(1)} + V^{(2)} c_{bulk}^{(2)} = 1. \tag{4.8}$$

We can assume the concentration of the free enzyme in region 1 and 2 is equal ($c_{bulk}^{(1)} = c_{bulk}^{(2)}$). Then

$$c_{bulk}^{(1)} \left[V^{(1)} + V^{(2)} + V^{(1)} x_{act} K_{ns}^0 l_{DNA}^{(1)} \right] = 1. \tag{4.9}$$

So the probability of being unbound $p_{unbound}$ is,

$$p_{unbound} = c_{bulk}^{(1)} (V^{(1)} + V^{(2)}) = \frac{V^{(1)} + V^{(2)}}{[V^{(2)} + V^{(1)}(1 + x_{act} K_{ns}^0 l_{DNA}^{(1)})]} \tag{4.10}$$

In this case its initial position is chosen randomly among all $N^{(1)} + N^{(2)}$ sites, and with probability x_{act} it is chosen to be in the active state (otherwise it is in the inactive state). If the walker is initially chosen to be bound, then it is placed randomly among the native DNA sites, and its state is initially set as active. When

4.2 Percolation cluster application — Modeling EcoRV's dynamic in E.coli cell

$x_{act} < 1$, it means EcoRV can switch between dormant and active states. The probability of switching from active to dormant state $p_{active \to dormant}$ is considered 0.20. Then we can define the probability of switching from dormant to active state $p_{dormant \to active}$ as $p_{dormant \to active}/p_{active \to dormant} = c_{active}/c_{dormant}$

$$p_{dormant \to active} = \frac{x_{act}}{1 - x_{act}} p_{active \to dormant}. \tag{4.11}$$

When $x_{act} = 1$, the enzyme is always active and cannot switch to dormant state. Therefore $p_{active \to dormant} = 1$ and $p_{dormant \to active} = 0$.

The other situation we should consider is when EcoRV is active and in region 1 (on native DNA). It will be bound non specifically and spends some time stuck on DNA. The probability of being stuck on DNA p_{stuck} can be obtained from Eq.(4.1)

$$p_{stuck} = \frac{c_{bound}^{(1)}}{c_{bound}^{(1)} + c_{act}^{(1)}} = K_{ns}^0 l_{DNA}^{(1)}/(1 + K_{ns}^0 l_{DNA}^{(1)}) \tag{4.12}$$

and the probability that EcoRV is released after one time step is $1 - p_{stuck}$. Then the probability of being released after nth time step is

$$\langle t \rangle = \sum_{n=1}^{\infty} n p_{stuck}^n (1 - p_{stuck}), \tag{4.13}$$

Therefore

$$\langle t \rangle = (1 - p_{stuck}) \sum_{n=0}^{\infty} (n' + 1) p_{stuck}^{n'}, \tag{4.14}$$

and

$$\langle t \rangle = (1 - p_{stuck}) \frac{d}{dp_{stuck}} \sum_{n=0}^{\infty} p_{stuck} = (1 - p_{stuck}) \frac{d}{dp_{stuck}} \frac{p_{stuck}}{1 - p_{stuck}}. \tag{4.15}$$

Finally we obtain

$$\langle t \rangle = \frac{1}{1 - p_{stuck}} = 1 + K_{ns}^0 l_{DNA}^{(1)} \tag{4.16}$$

Therefore EcoRV should spend a random time taken from the exponential distribution of $e^{-\langle t \rangle}$.

One should keep in mind that quantities $V^{(1)}$ is indeed fraction of available lattice sites on the cluster percolation in region 1 to the total number of sites in the corresponding region. The same goes for $V^{(2)}$ and $l_{DNA}^{(1)}$.

To convert the value of K_{ns}^0 we know that one base pair is $bp = 0.35$ nm $= 35 \times 10^{-10}$ m and one mole per liter is $6 \times 10^{23}/10^{-3} m^3$ and the binding constant $K_{ns}^0 = 10^7 M^{-1} bp^{-1}$, is $10^{-9}/(6.02 \times 3.5)$ m^2 and multiplication of $K_{ns}^0 l_{DNA}^{(1)}$ gives us a dimensionless quantity. The initial time is set to $t = 0$, and the search is carried out according to the following algorithm:

1. If the walker (EcoRV) is in the inactive state or is situated in the cytoplasm, a time τ_{step} is added to the total search time t. If the walker is active and situated in the region with the native DNA, then a random time is added to t. The random time it should spend on the native DNA is taken from an exponential distribution with average $\tau_{\text{step}}(1 + K_{\text{ns}}^0 l_{\text{DNA}}^{(1)})$.

2. One of the 6 directions possible on a cubic lattice is chosen at random. If a bond exists to a neighboring site in this direction, the walker moves to this site.

3. If $x_{\text{act}} = 0.01$ and the walker is in the active state, it switches with probability $1/5$ to the inactive state. If the walker is in the inactive state, it switches to the active state with probability $x_{\text{act}}/5/(1 - x_{\text{act}})$. In the case when $x_{\text{act}} = 1$, the walker always stays in the active state and switching is turn off.

4. If the walker is on the target site and in the active state, the target is considered to be found, and the time t is recorded as the search time. Otherwise the iteration goes back to step 1.

This procedure is repeated 5,000 times on the same percolation cluster, with a new target position and initial position of the searcher each time. The MFPT is calculated as the average of the 5,000 recorded search times.

4.2.1 Error bars

We can calculate the error bars by determining the standard deviation on N repetition.
The deviation is defined $\sigma(t) = \sqrt{\text{Var}(t)}$. The variance is

$$\text{Var}(t) = \sigma^2 = \sum_{i=1}^{N} \frac{t_i - \langle t \rangle}{N-1}, \qquad (4.17)$$

where t_i is the total search time of the ith repetition, and $\langle t \rangle$ is the MFPT. We have

$$\text{Var}(t_1 + t_2) = \text{Var}(t_1) + \text{Var}(t_2) \qquad (4.18)$$

and

$$\text{Var}(ct) = c^2 \text{Var}(t). \qquad (4.19)$$

Then

$$\text{Var}(\text{mean}) = \text{Var}(\sum_{i=1}^{N} t_i/N) = \frac{1}{N^2}\text{Var}(\sum_{i=1}^{N} t_i) = \frac{N}{N^2}\text{Var}(t) = \frac{1}{N}\text{Var}(t) \qquad (4.20)$$

Therefore the error bars have the value of mean deviation, $\sigma_{mean} = \sigma/\sqrt{(N)}$.

4.2.2 EcoRV's dynamic's modeled by percolation cluster

In Fig. 4.6 we compare the MFPT for EcoRV (activity $x_{act} = 0.01$) and mutant enzyme ($x_{act} = 1$) versus the bond occupation probability p. The bond occupation probability changes between the full occupation ($p = 1$, normal diffusion) down to the percolation threshold $p = 0.2488$ (subdiffusion with $\alpha = 0.51$). While the bonding constant K_{ns}^0 is $10^7 M^{-1} bp^{-1}$. For normal diffusion ($p = 1$) the MFPT is just a factor of two smaller for EcoRV, compared to the fully active mutant. Approaching the percolation threshold the native EcoRV increasingly outperforms the mutant, at criticality EcoRV's MFPT is *two orders of magnitude* shorter than that of the mutant. On average, the concentration of EcoRV is approximately

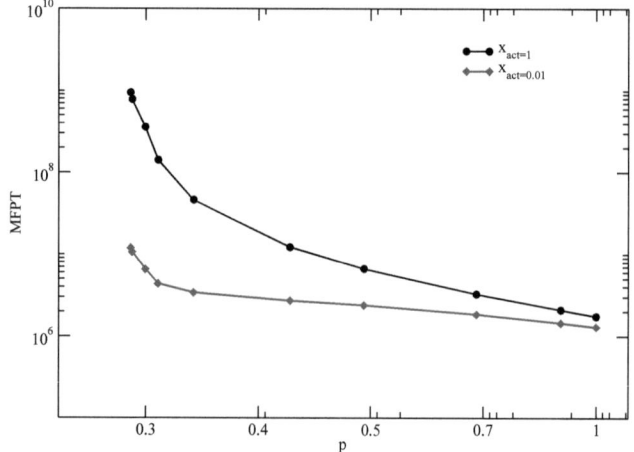

Figure 4.6: Typical time for the restriction enzyme to locate the target in an active state (MFPT) on a cubic lattice of size 100, as function of the bond occupation probability. Close to criticality ($p_c = 0.2488$), subdiffusion emerges with anomalous diffusion exponent $\alpha = 0.51$. The non-specific binding constant is $K_{ns}^0 = 10^7$ [$M^{-1} bp^{-1}$]. Error bars are of the size of the symbols or less.

$1/x_{act} = 100$ times higher in the cytoplasm outside the volume of the cell's native DNA than that of the mutant enzyme. At criticality, it is time-costly to cover distances, and thus EcoRV is 100 times more efficient than the fully active mutant enzymes. The mutant enzyme become trapped around the native DNA by they being bound non-specifically. In contrast, under normal diffusion conditions ($p = 1$) spatial separation is hardly significant, and the lower concentration is compensated by the higher activity of the mutant.

Even under severe anomalous diffusion with $\alpha \approx 0.51$ EcoRV's MFPT (for $x_{act} = 0.01$) is only a factor of ten higher than at normal diffusion. That means that the low-activity property of EcoRV renders their efficiency almost independent of the diffusion conditions, compared to the huge difference observed for the mutant.

Highly increased performance of the native EcoRV demonstrates that subdiffusion is not suppressing efficient molecular reactions. Also, our result show the reason for EcoRV's low-activity, that in this light appears as a designed property. It should be mentioned that the MFPT shown in our results is the result for an individual EcoRV enzyme. Typically, a bacteria cell combines a large number of restriction enzymes of different families. This significantly reduces the time scales shown here, while preserving the characteristics of the EcoRV superiority.

To obtain a better physical understanding of EcoRV non-specific binding to the cell's native DNA, we study the dependence of the MFPT on the non-specific binding constant K_{ns}^0. Experimentally, K_{ns}^0 can be varied by changing the salt concentration of the solution. For the cubic lattice ($p = 1$) we obtain an analytical expression for the MFPT for the geometry sketched in Fig. 4.4.

The simulations results for the case of normal diffusion are displayed in Fig. 4.7. At small K_{ns}^0, when the effect of native DNA is ignorable, the mutant ($x_{act} = 1$) clearly outperforms EcoRV, the gap in the MFPT corresponding to the reduced activity ($x_{act} = 0.01$), that is two orders of magnitude. As K_{ns}^0 increases both EcoRV and mutant perform almost identically, with a small (factor of two) advantage to EcoRV. For large K_{ns}^0 we can say that almost all active enzymes are bound to the cellular DNA, and EcoRV has approximately a factor of $1/x_{act} = 100$ higher bulk concentration. On the other hand its association rate constant with the target DNA in the cytoplasm is reduced by the same factor. In this normally diffusive regime dominated by non-specific binding, reduced activity of the restriction enzyme has no significant advantage. The resulting MFPT behavior according to Eq.(4.6) $T \approx K_{ns}^0 \ell_{DNA}^{(1)} V^{(1)}/k_0^a$ in this regime depends linearly on the non-specific binding constant. Indeed, this behavior is independent of x_{act}. The agreement between the theoretical model and the simulations results is excellent over the entire range of K_{ns}^0 (Fig. 4.7). Fig. 4.8 shows the behavior in the case of subdiffusion that is caused by crowding and in our model by diffusion on the percolation cluster. Almost for all values of K_{ns}^0, EcoRV significantly outperforms the mutant. At sufficiently large K_{ns}^0 values (above some 10^3 M^{-1}bp^{-1}) the value of the MFPT is approximately two orders of magnitude smaller, i.e., the performance is improved by a factor close to the value $1/x_{act}$. This behavior is thus dominated by the costly subdiffusion from the site of non-specific binding to the target. Because of subdiffusion on the percolation cluster, it takes more time for mutant enzyme, that almost all of them are bound

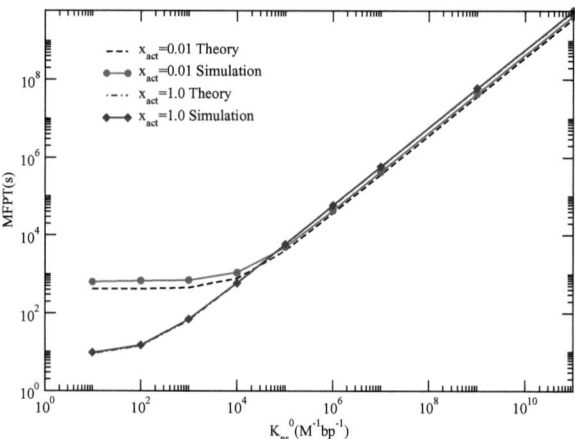

Figure 4.7: MFPT on a normal lattice ($p = 1$), as function of the non-specific binding constant K_{ns}^0. Simulations results are compared to the theoretical result Eq.(4.6). Comparison between the simulation and theoretical results shows a very good agreement.

non-specifically to the native DNA, to reach the target in cytoplasm. At low K_{ns}^0 values both curves converge. Now, the MFPT is fully dominated by anomalous diffusion to the target. Due to the compactness of the diffusion, that is caused by absence of some of bonds on the percolation cluster, the difference between EcoRV and the mutant becomes marginal and after an unsuccessful reaction attempt (when EcoRV hits the target but is in the dormant state), EcoRV has a higher chance to hit the target repeatedly before full escape, improving the efficiency. As we are dealing with random fractal, we have performed the simulation for different clusters and calculated the mean value. In Fig. 4.8 the thick lines show the average of simulations over three different critical percolation clusters, while the dotted lines display the result for each individual cluster. Apart from the low K_{ns}^0 limit, the results are very robust to the shape of the individual cluster. It would be interesting to derive analytically the MFPT dependence on K_{ns}^0. While the MFPT problem on a fractal has been solved recently [?], it is not clear how to apply this method in the present case, due to the division of the support into two subdomains. Similarly, for the related case of fractional Brownian motion [107] this remains an open question.

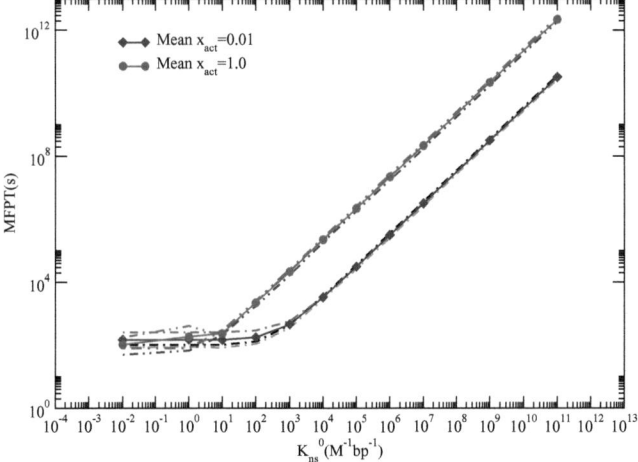

Figure 4.8: MFPT on a percolation cluster close to criticality ($p_c = 0.25$) versus the binding constant K_{ns}^0. The value $K_{ns}^0 = 10^7$ $M^{-1}bp^{-1}$ used in Fig. 4.6 is marked by the vertical line. Thick colored lines: average over three different percolation clusters. Dashed black lines: results for the individual clusters.

To obtain the mean first passage time for 1000 walkers, we choose a window of 1000 search times randomly among all of 10000 first passage times for a single walker we have and find the shortest search time (the fastest walker) among them that would be an equivalent of the first passage time for 1000 walkers. By repeating this process 100 times we can calculate the mean value of 100 shortest search times as the MFPT of 1000 walkers. In table 4.1 the result for 1000 walkers and single walker with different activities ($x_{act} = 0.01$ and $x_{act} = 1$) are presented. Behavior of the MFPT for 1000 walkers is similar to the the single walker that means the MFPT is shorter for EcoRV. In addition the concentration of enzymes has increased and we should expect a shorter the MFPT, considering that the ratio of the MFPT of mutant enzyme for 1000 walkers to the MFPT for EcoRV is larger than the single random walker case. Among 1000 mutant enzymes there will be more enzymes bound non-specifically to the native DNA while EcoRV has a higher concentration in cytoplasm.

It will be interesting to study the change in the MFPT by changing the number of

4.2 Percolation cluster application — Modeling EcoRV's dynamic in E.coli cell

Walkers	MFPT(s) ($x_{act} = 1$)	σ	MFPT (s) ($x_{act} = 0.01$)	σ
1	5.89×10^5	5.88×10^3	4.03×10^5	4.28×10^3
1000	5.05×10^2	16.1	2.85×10^2	9.2

Table 4.1: The MFPT for one random walker and 1000 random walkers when $K_{ns}^0 = 10^7 M^{-1} bp^{-1}$ on regular lattice.

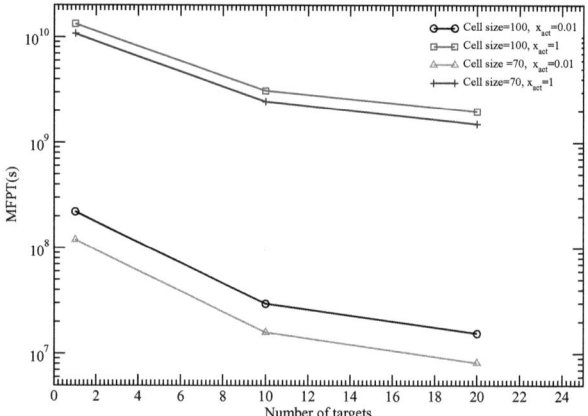

Figure 4.9: The MFPT versus the number of targets. Number of targets vary between one to 20, each target occupies a lattice site. and $K_{ns}^0 = 10^7 M^{-1} bp^{-1}$, on percolation cluster ($p = 0.25$) on a lattice of size 70 by 70 by 70 and 100 by 100 by 100 lattice sites.

targets and volume of the cytoplasm. The simulation has been executed for target size of one, 10 and 20 lattice sites, and $K_{ns}^0 = 10^7 M^{-1} bp^{-1}$, on the percolation cluster on a lattice of size 70 by 70 by 70 and 100 by 100 by 100 lattice sites (Fig. 4.9). All targets are places on one straight line, and the line is placed randomly in cytoplasm for each realization. In all cases probability of hitting the target is increased and the MFPT decreases by increasing the number of targets. By changing the lattice size while keeping the ratio of DNA and cell radius ($R_{DNA}/R_{cell} = 0.5$), there will be less sites for EcoRV to explore therefore the MFPT is shorter for both kind of enzymes. This effect is more obvious for EcoRV because it has a higher concentration in cytoplasm than the mutant enzyme and reaches the target faster.

4.2 Percolation cluster application — Modeling EcoRV's dynamic in E.coli cell

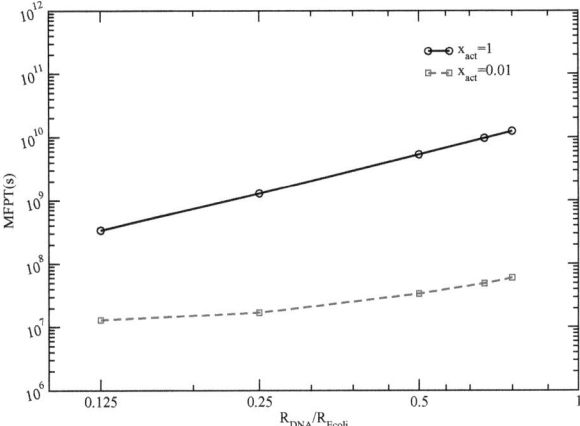

Figure 4.10: The MFPT of two enzymes versus R_{DNA}/R_{cell} on percolation cluster ($p = 0.25$), $K_{ns}^0 = 10^7 M^{-1} bp^{-1}$. The MFPT of both EcoRV and mutant enzyme increase by increasing R_{DNA}/R_{cell}, but the change for EcoRV is not as rapid as mutant enzyme.

We studied the effect of changing the ratio R_{DNA}/R_{cell} on the search efficiency. R_{DNA}/R_{cell} varies between 1/8 to 3/4 and the MFPT of both EcoRV and mutant enzyme increase by increasing R_{DNA}/R_{cell}, as there will be larger volume of DNA for both enzymes to get trapped in it. In Fig. 4.10, the MFPT of two enzymes versus R_{DNA}/R_{cell} is displayed. The change in the MFPT of EcoRV is not as much as mutant's MFPT, because although the volume of native DNA and therefore trapping sites is increasing, EcoRV is still to switch to dormant state and the MFPT does not increase as rapid as permanently-active mutants that spends more time on DNA as DNA volume increases.

In Fig. 4.11 we compare the MFPT for EcoRV (activity $x_{act} = 0.01$) and mutant enzyme ($x_{act} = 1$) versus the bond occupation probability p with different initial conditions. The bond occupation probability varies between the full occupation ($p = 1$, normal diffusion) down to the percolation threshold $p = 0.2488$ (subdiffusion with $\alpha = 0.51$). Bonding constant K_{ns}^0 is $10^7 M^{-1} bp^{-1}$. Once enzymes (EcoRV and mutant) choose their starting sites among the native DNA sites. The other initial condition would be when the enzymes starting sites are chosen only among cytoplasm sites. The final result does not depend on the initial conditions and the behavior is

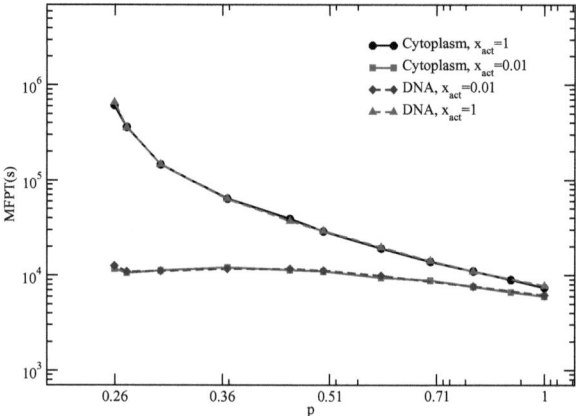

Figure 4.11: The MFPT on a cubic lattice of the size 70 by 70 by 70 (smaller than the 100 by 100 by 100 lattice in Fig.4.6), as function of the bond occupation probability for different initial conditions. Close to criticality ($p_c = 0.2488$), subdiffusion emerges with anomalous diffusion exponent $\alpha = 0.51$. The non-specific binding constant is $K_{ns}^0 = 10^7$ [M^{-1}bp^{-1}]. Cytoplasm and DNA in the figure, indicate the initial condition for EcoRV ($x_{act} = 0.01$) and the mutant enzyme ($x_{act} = 1$).

similar to Fig. 4.6. In our simulations we had in mind biologically relevant situation, namely that the walker on the average is placed in an equilibrated fashion. This means that the walker has been around for a sufficiently long time such that its position is random in the cell. Therefore an average over initial positions appears as the appropriate choice. For the restriction enzyme this assumption appears justified. These enzymes are contained in the cells at any time and typical degradation times surpass the life time the cell.

The First passage time density of both enzymes with different initial conditions (starting just on the native DNA or cytoplasm) and bond probability $p = 0.3$ is displayed in Fig. 4.12a. Because of the finite space FPTD has an exponential tail that gives the characteristic time (or the MFPT). We see that although the shortest time range belong to mutant enzyme that is starting in cytoplasm, longest search times belong to mutant enzyme $x_{act} = 1$ with both initial conditions on native DNA and cytoplasm, and the order of longest time is much larger than shorter search times that influence the final MFPT of the enzyme.

4.2 Percolation cluster application   Modeling EcoRV's dynamic in E.coli cell

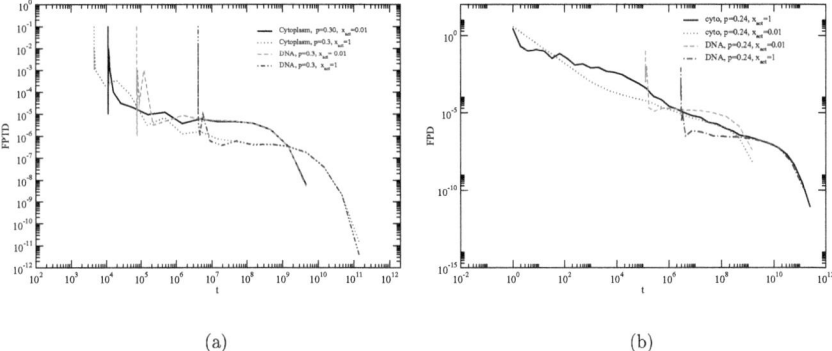

(a)   (b)

Figure 4.12: (a) The FPTD for EcoRV ($x_{\text{act}} = 0.01$) and the mutant enzyme ($x_{\text{act}} = 1$) with different initial conditions on percolation cluster with bond probability $p = 0.3$, $K_{\text{ns}}^0 = 10^7 [\text{M}^{-1}\text{bp}^{-1}]$. (b) The First passage time density of both enzymes with different initial conditions with bond probability $p = 0.2488$ and $K_{\text{ns}}^0 = 10^7 [\text{M}^{-1}\text{bp}^{-1}]$.

In Fig. 4.12b first passage time density of both enzymes with different initial conditions with bond probability $p = 0.2488$ is displayed. Although the shortest time range belong to mutant enzyme that is starting in cytoplasm, longest search times belong to mutant enzyme $x_{\text{act}} = 1$ with both initial conditions on native DNA and cytoplasm, and the order of longest time is much larger than shorter search times that influence the final the MFPT of the enzyme.

The FPTD for EcoRV ($x_{\text{act}} = 0.01$) and the mutant enzyme ($x_{\text{act}} = 1$) with bond probability $p = 1$ and different lattice sizes (50,150,200), without native DNA in the center ($K_{\text{ns}}^0 = 0 [\text{M}^{-1}\text{bp}^{-1}]$) is presented in Fig. 4.13a. As it is expected in the subdiffusion regime the range of first passage time for the mutant enzyme increases. In contrary with the subdiffusive regime, in this case the shortest first passage times belong to mutant enzyme that outperforms EcoRV in absence of the native DNA. By increasing the lattice size gap between the shortest FPT for mutant and EcoRV is considerable.

In Fig. 4.13b FPTD of EcoRV and mutant in the regular lattice of size 20, without native DNA is displayed. The target is a cube of size 2 fixed in the center. We see that because of absence of DNA the mutant's search is faster. In Fig. 4.14 the MFPT as a function of bond probability for a cell without native DNA is displayed. Even in subdiffusion regime, the mutant enzyme outperforms EcoRV. In out theoretical results, we expect the same behavior in regular lattice and in the subdiffusive regime the difference is more obvious.

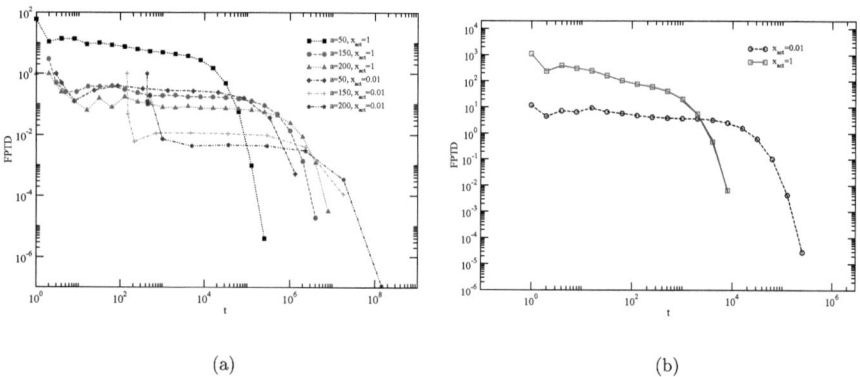

Figure 4.13: (a) The FPTD for EcoRV ($x_{act} = 0.01$) and the mutant enzyme ($x_{act} = 1$) with bond probability $p = 1$ and different lattice sizes (50,150,200), without native DNA in the center ($K_{ns}^0 = 0[\text{M}^{-1}\text{bp}^{-1}]$) (b) The FPTD for EcoRV ($x_{act} = 0.01$) and the mutant enzyme ($x_{act} = 1$) with different lattice sizes in regular lattice $p = 1$, $K_{ns}^0 = 0$ [$\text{M}^{-1}\text{bp}^{-1}$].

4.3 Fractional Brownian motion application

Another alternative approach to this problem is fractional Brownian motion. Fractional Brownian motion is not only of interest for communications engineers, it has applications in other areas, such as finance, physics, probability, statistics, hydrology, biology and bioengineering. In bioengineering for instance, fractional Brownian motion is used to model regional flow distributions in the heart, the lung and the kidney [108] and viscolastic properties of polymers in solutions [109].

The choice of a non-Markovian process seemed appropriate as the experimentally observed subdiffusion is claimed to be a consequence of the viscoelasticity of the intracellular fluids [110]. Particles in cytoplasm and nucleoplasm stay longer at their original position and return slower when having escaped to a far-away distance and the poor spreading associated with subdiffusion implies a slow sampling process. Subdiffusion in a viscoelastic fluid is modeled via fBm using Hosking method [111] as an exact method or Weierstrass-Mandelbrot function (WMF) to generate fractional gaussian noise [23].

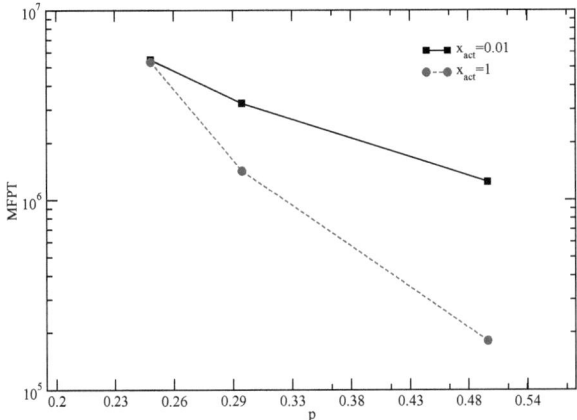

Figure 4.14: The MFPT as a function of bond probability for a cell without native DNA is displayed. Even in subdiffusion regime, the mutant enzyme outperforms EcoRV.

4.3.1 Generating fractional Brownian motion

From Eq.(2.51) we see that if $H = 1/2$, all the covariances are zero and the fractional Gaussian noise $(fGn(k) = B_H(k+1) - B_H(k))$ is a Gaussian process that implies independence and agrees with the properties of Brownian motion, which has independent increments. Covariances are negative for $H < 1/2$ and fBm is anti-persistence. Covariances are positive for $H > 1/2$ and fBm is persistence. For $H = 0.3$, the negative correlation accounts for the high variability, on the other hand for $H = 0.7$ there are clearly periods in which the sample path increases and periods in which it decreases. In Fig. 4.15, fBm path samples generated by Hosking method are displayed. The negative correlations for $H = 0.3$ are also observed in this plot, whereas the sample is more smooth for $H = 0.7$ due to the positive correlations.

The Hosking method

The Hosking method is an algorithm to simulate a general stationary Gaussian process. We will focus on the simulation of fractional Gaussian noise (fGn) $fGn(0), fGn(1)...$. Fractional Brownian motion sample is obtained from a fractional Gaussian noise

4.3 FBm application

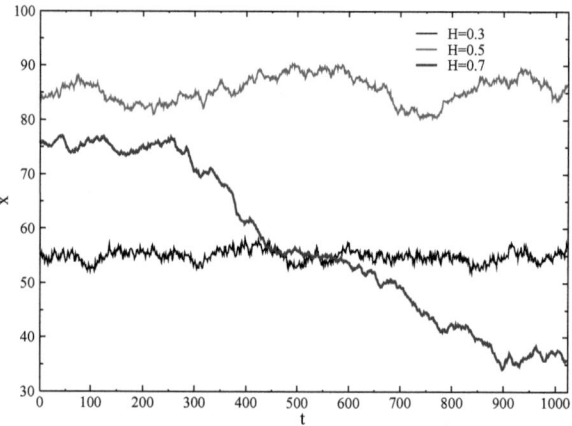

Figure 4.15: Trajectory of a random walker in one dimension with Hurst parameters $H = 0.3, 0.5, 0.7$. The correlations are negative for $H = 0.3$ and the sample is more smooth for $H = 0.7$ due to the positive correlations.

sample by taking cumulative sums. The method generates X_{n+1} given $X_n, ...X_0$ recursively. It does not use specific properties of fractional Brownian motion nor fractional Gaussian noise. The algorithm can be applied to any stationary Gaussian process. The key observation is that the distribution of X_{n+1} given the past can explicitly be computed. We used this method which is an exact method to generate fractional Gaussian noise (fGn). Later we used Weierstrass-Mandelbrot function to generate the fGn in shorter time. Therefore we do not explain this method in detail and more explanation of Hosking method is in [116]. The code using Hosking method to generate fGn is attached in Appendix B.

Weierstrass-Mandelbrot function (WMF)

To mimic the subdiffusion of particles in crowded intracellular fluids like the cytoplasm, we have determined the diffusive steps according to the Weierstrass-Mandelbrot function (WMF), see Eq.(4.21). The WMF yields a path with the characteristics of fractional Brownian motion [113, 114], i.e., the individual step sizes are not independent but correlated. For instance the WMF models the fluid's memory that is reflected in a nontrivial creep function [115]. We started N individual, noninteracting particles 1D, 2D, and 3D to model the EcoRV subdiffusion in E.coli's cell. They followed their (subdiffusive) random walk up to a time T_{max}. The erratic motion of the particles was simulated using the forward integration of the Langevin equation, i.e., the positions at times $t = 1, 2, ..., T_{max}$ were obtained by $x_i(t+1) = x_i(t) + \xi_i$ with $i = 1, 2, 3$. As a model for subdiffusive motion we have chosen to calculate the spatial increments $\xi_i(t) = fGn_i(t) = W(t+1) - W(t)$ in each spatial direction $i = 1, 2, 3$ by the Weierstrass-Mandelbrot function [113, 114]:

$$W(t) = \sum_{n=-\infty}^{\infty} \frac{\cos(\phi_n) - \cos(\gamma^n t + \phi_n)}{\gamma^{n\alpha/2}} \quad (4.21)$$

Here, ϕ_n are random phases in the interval $[0, 2\pi]$, $\gamma > 1$ is an irrational number, T_{max} is the length of the desired time series, and α is the degree of anomaly that appears in the MSD ($\langle r^2(t) \rangle \sim t^\alpha$). In accordance with Saxton [114], we have chosen $\gamma = \sqrt{\pi}$ and restricted the sum to the terms $n = -8, ...48$. The errors caused by this limitations are negligible [23, 114].

4.3.2 Mean squared displacement for fBm

The MSD of a fBm process in one dimensional infinite space with different Hurst parameters 0.3, 0.5 and 0.7 is displayed in Fig. 4.16a. Mean squared displacement is $\langle r^2(t) \rangle \sim t^{2H}$. All walkers start at $x(t_0) = 0$, and fGn for T_{max} steps is generated with the help of Hosking method. The next step is defined as $x(t_i) = x(t_{i-1}) + fGn(t_i)$. Fig. 4.16b shows the Time averaged MSD and the ensemble averaged MSD of fBm in one dimensional finite space with Hurst parameters $H = 0.3$. Due to the diffusing in the finite space ($L = 2$) the ensemble averaged MSD follows Eq.(2.64) and reaches a plateau of value $L^2/3$ and the value of the plateau for the TA MSD is $2L^2/3$. This difference comes from the definition of TA MSD that follows Eq.(2.67). Both MSDs have the same slopes that shows the ergodicity of fBm. Fig. 4.17a depicts the behavior of the MFPT for fBM process in one dimension as a function of Hurst parameter. The MFPT increases as $H \to 0.5$. It was mentioned previously (Eq.(2.62)) that the first passage time in semi infinite domain is $p_{fp}(t) \sim t^{H-2}$, therefore the MFPT increases by increasing the value of H. The random walker starts at $x = 0$, in the middle of the one dimensional box, and the target is at $x = 1.5$. At $x = -1.5$ we have reflecting boundary conditions.

4.3.3 EcoRV's dynamic's modeled by fBm

We considered the same initial conditions and theory for EcoRV (see section 4.2) to study this problem with fBm process. The enzyme searches for DNA target with all probability and parameters explained in section 4.2, only not in the lattice but in the continuous space. In addition reflective boundary conditions have been applied

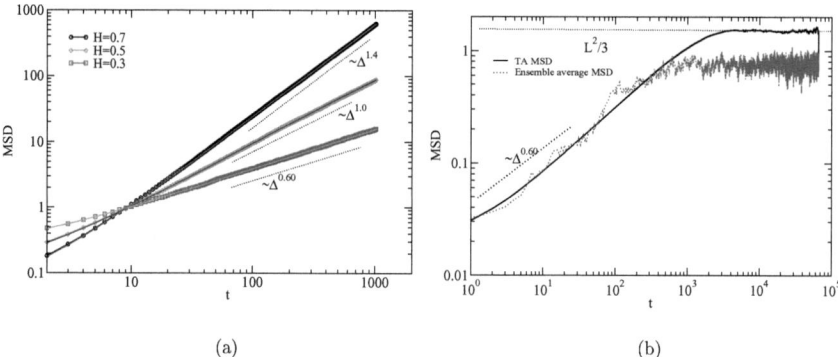

Figure 4.16: (a) The MSD of fBm in one dimensional infinite space with different Hurst parameters 0.3, 0.5 and 0.7. MSD is $\langle r^2(t) \rangle \sim t^{2H}$. (b) The time averaged MSD and the ensemble averaged MSD of fBm in one dimensional finite space with different Hurst parameters 0.3. MSDs are proportional to t^{2H} and due to diffusion in finite space reach a plateau.

for the cell's walls in three dimensions. For example if during the simulations the random walker jumps in x direction behind the walls ($|x(t)| > L$ at some time t is bounced back to the position $x(t) - 2|x(t) - \text{Sign}(x)L|$.

Fig. 4.18 depicts the MFPT of two enzymes as a function of Hurst parameter for EcoRV $x_{act} = 0.01$ and mutant enzyme $x_{act} = 1$ and the binding constant is $K_{ns}^0 = 10^7 M^{-1} bp^{-1}$. For all values of $H < 1/2$ we are in subdiffusive regime and the MFPT is increasing by as $H \to 1/2$ (see Fig. 4.17 to compare the results in one dimension). In the subdiffusive regime the value of the MFPT is always larger for the mutant enzyme $x_{act} = 1$. We know from our theoretical results for normal diffusion case (Eq.(4.6)) and diffusion on the regular lattice ($p = 1$ in Fig.4.6)how the results will be when $H = 0.5$. In limit of $H \to 1/2$ the necessary computing time exceeds the capability of our work stations. Therefore this limit is not the part of the following. In percolation cluster anomaly is $2/d_w = 0.51$ therefore in our simulation for EcoRV, we have chosen $\alpha = 0.5$ (or $H = 0.25$) to compare the effects with percolation cluster results. Dependency of the MFPT to the binding constant for EcoRV $x_{act} = 0.01$ and mutant enzyme $x_{act} = 1$ performing fBm with $H = 0.25$ is displayed in Fig. 4.19. Almost for all values of K_{ns}^0, EcoRV significantly outperforms the mutant. At large K_{ns}^0 values (above some 10^3 M^{-1}bp^{-1}) the value of the MFPT is approximately two orders of magnitude smaller, i.e., the performance is improved by a factor close to the value $1/x_{act}$. At low K_{ns}^0 values both curves converge. The behavior is very

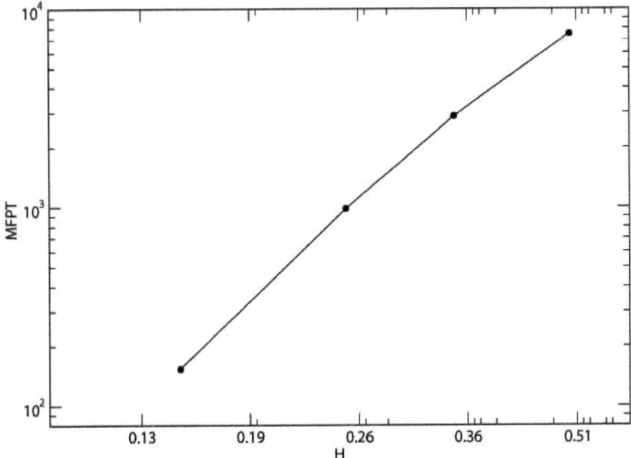

Figure 4.17: The MFPT for fBm in one dimension as a function of Hurst parameter. The random walker starts at $x = 0$ and the target is at $x = 1$. The FPTD in semi infinite domain is $p_{fp}(t) \sim t^{H-2}$, therefore the MFPT increases by increasing the value of H.

similar to the diffusion on the percolation cluster. The FPTD has a finite mean and exponential tail for the fractal (percolation)[97] and fBm [49] model in finite domain, while it has an infinite mean and a power law tail in a CTRW model. And also both processes have increased probability to immediately return to points just visited. Similarity of fBm and diffusion on fractals is the topic of ongoing research [117].

4.4 Continuous Time Random Walk application

The CTRW induces subdiffusion by altering the timing between two diffusional steps yielding a diffusion equation with a fractional time derivative. The overall density of free proteins and molecular aggregates is very high in the cytoplasm. In this crowded environment, the tracer particle is trapped in dynamic cages and their life

4.4 CTRW application

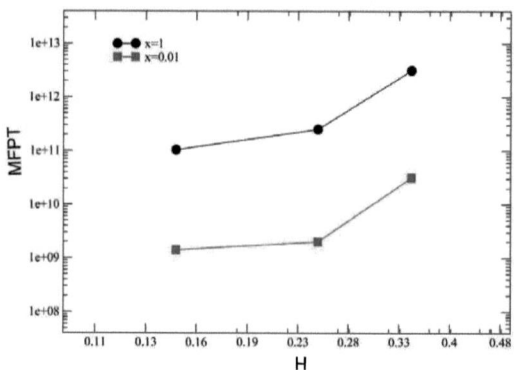

Figure 4.18: The MFPT for fbm as a function of Hurst parameter H for EcoRV $x_{act} = 0.01$ and mutant enzyme $x_{act} = 1$ for $K_{ns}^0 = 10^7 M^{-1} bp^{-1}$. Similar to one dimensional fBm the MFPT increases by increasing H.

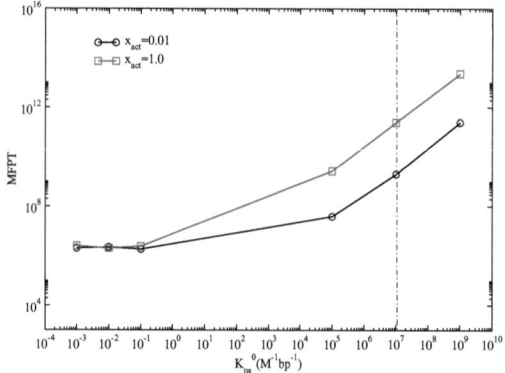

Figure 4.19: The MFPT for fbm as a function of binding constant for EcoRV $x_{act} = 0.01$ and mutant enzyme $x_{act} = 1$ for $H = 0.25$. fBm and percoaltion clusters share the same nature and the their results should be similar.

times are broadly distributed at high densities and leading to

$$\psi(t) = \frac{\alpha/\tau}{(1+t/\tau)^{1+\alpha}}. \tag{4.22}$$

Therefore we use CTRW with heavy-tailed waiting time distribution that has an infinite MFPT ($\langle t \rangle \to \infty$) to study our case. Eq.(4.22) asymptotically has the power law behavior $\psi \simeq \tau^\alpha/t^{1+\alpha}$ and converges for $t \to 0$ and is normalized.

4.4.1 Generating waiting times

The random walker takes steps on the lattice and in each step it waits a random time taken from Eq.(4.22). To generate a random number with the distribution given by Eq.(4.22), a commonly used technique is called the inverse transform technique. Let y be a uniform random variable from the interval [0,1]. If $X = F^{-1}(y)$, then X is a random variable with the cumulative distribution function $F_X(x) = F$.

$$F(t) = \int_{t_0}^{t} \psi(t')dt' \equiv y. \tag{4.23}$$

Therefore, if $\tau = 1$ from Eq.(4.22) we obtain,

$$t = (1-y)^{-1/\alpha} - 1, \tag{4.24}$$

t is the random variable with a distribution $\psi(t)$.

In each step the random walker chooses a direction but the time is counted as an integer number of random variable t. Fig. 4.20a shows trajectories of random walkers performing CTRW on one dimensional lattice with different values of the anomalous coefficient α for 100 time steps. Random walkers starting position is chosen randomly between [0,80]. As the value of α decreases the random walker should wait a longer time and for $\alpha = 0.2$ and $\alpha = 0.5$ we see an extreme stalling.

Fig. 4.20b displays the ensemble averaged MSD and the ensemble averaged time averaged MSD and the time averaged MSD for individual trajectories in one dimension. They exhibit a considerable scatter in amplitude around the ensemble averaged TA MSD. According to Eq.(2.70) the TA MSD is linear in time and the ensemble averaged MSD is proportional to t^α. In Fig. 4.21a mean squared displacement for CTRW in three dimensions with $\alpha = 0.25$ is displayed. Similar to the one dimensional case the TA MSD is linear in time and the ensemble averaged MSD is proportional to $t^{0.25}$. The FPTD of CTRW process in one dimensional lattice is displayed in Fig. 4.22a. According to Eq.(2.61) the first passage time density in a finite box is proportional to $t^{-1-\alpha}$. Here $\alpha = 0.5$ therefore FPTD is a power law with slope of -1.5. Using logarithmic binning we obtain a very good result. Fig. 4.22b shows the FPTD for CTRW with $\alpha = 0.25$ in one dimensional finite box.

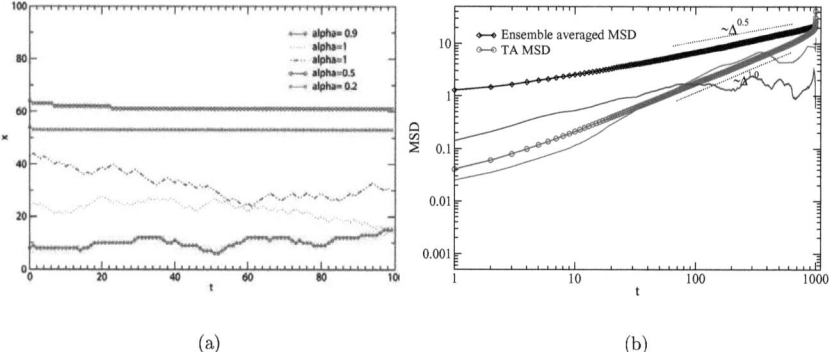

Figure 4.20: (a) Trajectories of random walkers performing CTRW on one dimensional lattice with different values of anomalous diffusion coefficient α for 100 time steps is shown. Random walkers start position is chosen randomly between [0,80]. CTRW with $\alpha < 0.5$ causes an extreme stalling. (b) Ensemble averaged MSD and ensemble averaged time average MSD and time average MSD for individual trajectories for CTRW with $\alpha = 0.5$ in one dimension ensemble averaged MSD and ensemble averaged time average MSD and time average MSD for individual trajectories for CTRW with $\alpha = 0.5$ in one dimension

4.4.2 Waiting time distribution with a cutoff

In finite systems, cutoffs occur naturally corresponding to, e.g., a maximal well depth in a random energy landscape. We choose a power law waiting time distribution with a cutoff and introduce the cutoff τ^* effective at $t \ll \tau^*$.

$$\psi(t) = \frac{d}{dt}\left[1 - \frac{\tau^\alpha}{t+\tau^\alpha}e^{-t/\tau^*}\right]. \qquad (4.25)$$

In free space Eq.4.25 produces initial subdiffusion $\langle r^2(t)\rangle \simeq t^\alpha$, and turns over to $\langle r^{(}t)\rangle \simeq t$ at $t \ll \tau^*$. To generate a random number from Eq.(4.25) distribution we should follow the same instruction to generate a random number with power law distribution by obtaining the cumulative function $F(t)$ and calculating $F(t)-y=0$, where y is the random variable with uniform distribution. Here calculating t is not as easy as Eq.(4.24) but we can solve the equation by using numerical methods. In appendix C the code to solve $F(t) - y = 0$ and generation of the random number according to Eq.(4.25) distribution, is attached.

4.4 CTRW application — Modeling EcoRV's dynamic in E.coli cell

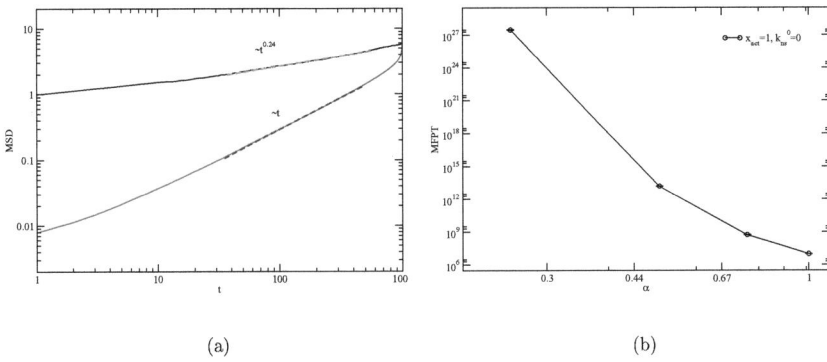

Figure 4.21: (a) The Mean squared displacement for CTRW in 3 dimension with $\alpha = 0.25$. The TA MSD is linear in time while in the ensemble averaged MSD is proportional to $t^{0.25}$ (b) MFPT versus the anomaly α in CTRW.

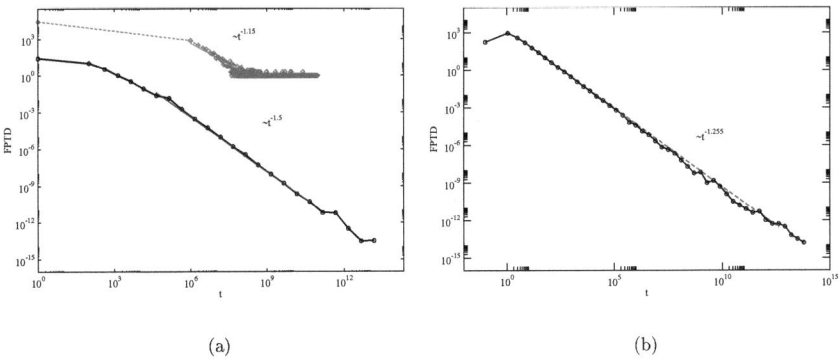

Figure 4.22: (a) The FPTD of CTRW process in one dimension with $\alpha = 0.5$. (b) The FPTD of CTRW process in one dimension with $\alpha = 0.25$. Both FPTDs are power law $t^{-1-\alpha}$.

4.4.3 EcoRV's dynamic's modeled by CTRW

We study EcoRV's diffusion performing continuous time random walk in three dimensional lattice with the same sizes and probabilities defined in section 4.2.

4.4 CTRW application Modeling EcoRV's dynamic in E.coli cell

Fig. 4.23 depicts the dependency of the MFPT on binding constant $K_{ns}^0[\text{M}^{-1}\text{bp}^{-1}]$ with anomaly similar to percolation and fBm case $\alpha = 0.5$ for the mutant enzyme $x_{act} = 1$ and EcoRV $x_{act} = 0.01$. EcoRV diffuses in finite space therefore choosing a waiting time distribution with a cutoff [19] makes the simulation more realistic. In random walk on the regular lattice case, the MFPT is around $10^4 - 10^5$ time steps therefore we assume after 10^4 steps the random walker has found the target and choose the cutoff time $\tau^* = 10000$ time steps. In this case the mutant enzyme outperforms EcoRV for small values of binding constant but for larger values of K_{ns}^0 there is not an obvious different on enzymes performance and EcoRV's performance is slightly more efficient. The results for this case are very similar to the case for normal random walk on regular lattice (Fig.4.7). It seems unlike percolation and fBm case, the subdiffusive behavior of EcoRV under a heavy tailed waiting time distribution does not improve the search efficiency of EcoRV.

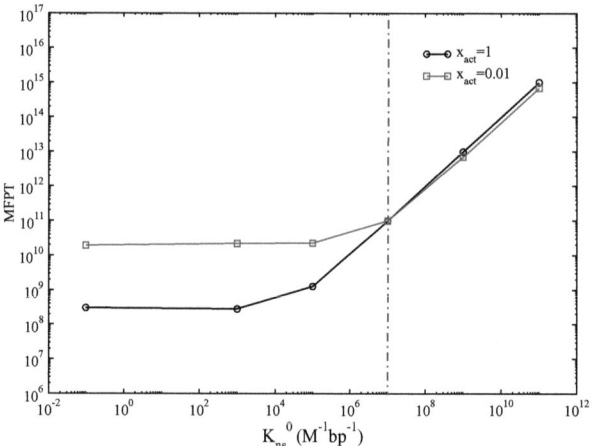

Figure 4.23: The MFPT versus $K_{ns}^0[\text{M}^{-1}\text{bp}^{-1}]$ for CTRW with $\alpha = 0.5$ with power law waiting time distribution with cutoff ($\tau^* = 10000$) on regular lattice. Two cases are studies. (a) Mutant enzyme $x_{act} = 1$ (b) EcoRV which switches between dormant and active state while moving to another lattice site $x_{act} = 0.01$.

4.4.4 CTRW on percolation cluster

Previously we discussed diffusion on percolation cluster and CTRW that are two different models for subdiffusion in cytoplasm and each leading to subdiffusion. Both a CTRW and diffusion on a fractal have non-Gaussian propagators. Different methods such as p variation tests for categorizing diffusion are emerging [118, 119]. Currently these tests are still too simple and, both mechanisms can coexist [41]. Few studies started to consider both mechanism to study the anomalous diffusion in the cell [40]. We consider the synergy of a non ergodic heavy-tailed CTRW on a fractal and study our system when the particle moves in percolation clusters as we described before and also waits for a random time (taken from a power law waiting time distribution) on each site of the lattice. The ensemble averaged and ensemble

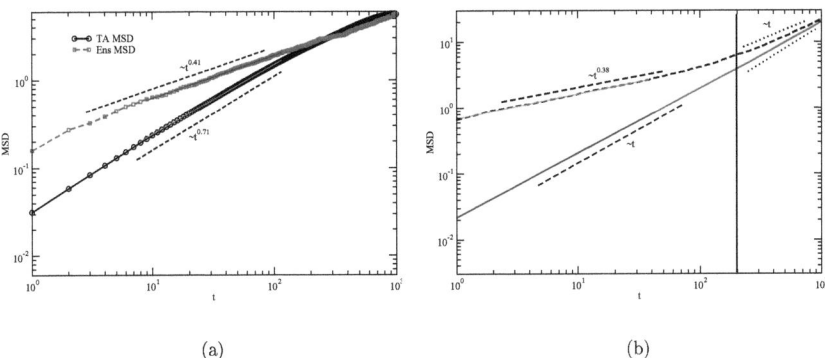

(a) (b)

Figure 4.24: (a) The ensemble averaged and the ensemble averaged time averaged mean squared displacement of CTRW with $\alpha = 0.7$ on percolation cluster in 3 dimension on cubic lattice with $p_c = 0.25$. (b) Ensemble averaged and ensemble averaged time averaged mean squared displacement for CTRW with $\alpha = 0.4$ on regular cubic lattice $p_c = 1$. Cutoff time $\tau^* = 200$ time steps and the MSD is linear in time for time steps $t > \tau^*$.

averaged time averaged mean squared displacement for CTRW with $\alpha = 0.7$ on percolation cluster in three dimensions on cubic lattice with $p_c = 0.25$ are displayed in Fig. 4.24a. According to Eqs.(2.71) and (2.72) for percolation cluster, anomaly is $\beta = 2/d_w = 0.51$ and the ensemble averaged MSD in this case should be proportional to $t^{\alpha\beta} = t^{0.36}$ and the TA MSD should be $t^{1-\alpha+\alpha\beta} = t^{0.3+0.36}$. Fig. 4.24b shows ensemble averaged and ensemble averaged time averaged mean squared displacement for CTRW with $\alpha = 0.4$ on regular cubic lattice $p_c = 1$. Cutoff time $\tau^* = 200$ and the MSD is linear in time for time steps $t > \tau^*$. For $t < \tau^*$ the ensemble averaged MSD is proportional to $t^{\alpha\beta} = t^{0.4}$ ($\beta = 1$) and the TA MSD is linear in time $t^{1-\alpha+\alpha\beta} = t$. We applied CTRW on percolation to model EcoRV's diffusion in E.coli

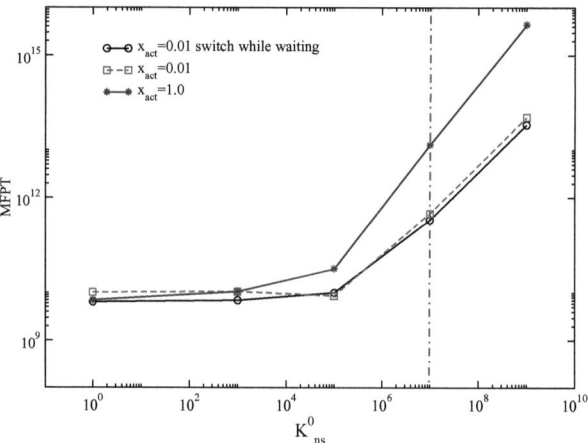

Figure 4.25: The MFPT versus $K_{ns}^0 [\text{M}^{-1}\text{bp}^{-1}]$ on percolation percolation cluster $p_c = 0.25$ and performing CTRW with $\alpha = 0.5$. Three cases are studies. (a) Mutant enzyme $x_{act} = 1$ (b) EcoRV which switches between dormant and active state while moving to another lattice site $x_{act} = 0.01$. (c) EcoRV which switches while moving to another lattice site $x_{act} = 0.01$ and also while it is waiting due to CTRW process.

cell. Fig. 4.25 depicts the dependency of the MFPT on binding constant K_{ns}^0. The result is very similar to the percolation (Fig. 4.8) and fBm (Fig. 4.19) results and EcoRV outperforms the mutant enzyme for all values of binding constant. The CTRW on percolation for ($\alpha \leq 1$) seems to be dominated by the fractal nature. We also considered the situation that EcoRV switches between dormant and active state while it is trapped on a lattice due to a power law waiting time distribution of CTRW. The mean first passage time for this case is always smaller than the case does not switch while waiting but does not show a huge advantage.

Conclusions

We presented a case study of EcoRV restriction enzymes involved in vital cellular defence. EcoRV belongs to the range of sizes for which subdiffusion under crowding was reported [15, 20]. It is often argued that molecular processes in the cell could not be subdiffusive, as this would compromise the overall fitness of the cell due to the slowness of the response to external and internal perturbations.

We analyzed different stochastic processes for subdiffusion and demonstrated a solution to this subdiffusion-efficiency paradox. Specific molecular design renders the efficiency of EcoRV enzymes almost independent on the exact diffusion conditions. This effect is more pronounced in percolation and fBm process. Even though EcoRV is not always ready to bind, under subdiffusion conditions the low enzyme activity represents a superior strategy. We showed that due to its so far elusive propensity to an inactive state the enzyme avoids non-specific binding and remains well-distributed in the bulk cytoplasm of the cell. Despite the reduced volume exploration of subdiffusion processes, the low activity of the enzyme surprisingly guarantees a high efficiency of the enzyme.

The cytoskeleton resembles a fractal and thus can be modeled by random walks on critical percolation cluster to mimic this subdiffusion. Cellular subdiffusion may also be modeled by fractional Brownian motion (FBM) or continuous time random walks (CTRW) [120]. FBM shares many features with diffusion on fractal structures, e.g., the compactness and ergodicity. In fBm process, particle's exploration is compact as its walk dimension is given by $d_w = 2/\alpha$ which $\alpha = 2H$ [38] and similar to the percolation cluster in case of d dimensional diffusion (d>2) the walk dimension in larger than the space dimension. Because of this compact exploration crowding-induced subdiffusion does not hamper the cell but can be used to enhance the cell performance. A strong anomaly (low α) leads to an increasing probability of finding the target. Therefore we expect that the essential observations for the MFPT found for percolation cluster should be similar for the case of FBM (Fig. 4.26). In CTRW similar to normal Brownian motion the walk dimension is $d_w = 2$ and the random walker explores a surface completely but will only visit a negligible subspace when moving in three-dimensional bulk solution. For CTRW, subdiffusion is induced by not moving in a given period of time (because of power law waiting time distribution). Therefore the results in case study of EcoRV may be similar to the Brownian motion We considered the possibility of coexistence of two different

4.4 CTRW application

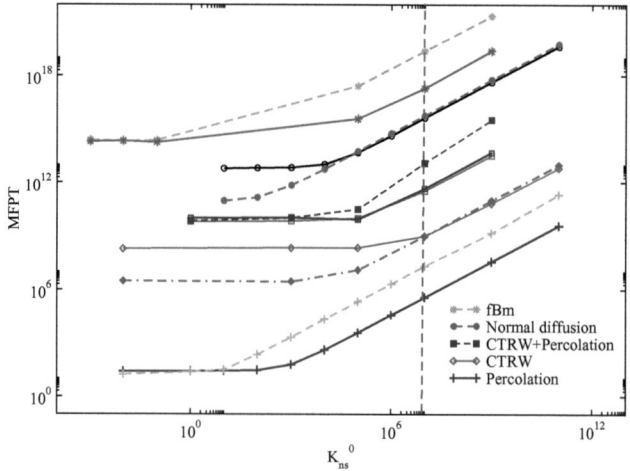

Figure 4.26: The MFPT versus the binding constant for different processes, diffusion on regular lattice ($p = 1$), diffusion on critical percolation ($p = 0.25$), fBm, CTRW on regular lattice and CTRW on percolation cluster. The vertical line indicates $K_{ns}^0 = 10^7$ M^{-1}bp^{-1}. The MFPT for mutant enzyme are shown by dashed line.

processes such as CTRW and diffusion on fractals and thus performed our diffusion for CTRW on percolation. We studied the case in which EcoRV while being trapped next to the target EcoRV would have ample chance to convert to the active state and knock out the target.

Our results demonstrate that reduced non-specific binding are beneficial for efficient subdiffusive enzyme activity even in relatively small bacteria cells. It seems intracellular fluids have just the right amount of crowding to induce an anomalous diffusion near to the critical α [16, 20, 110]. We compare the results for different processes in Fig. 4.26. The values of MFPT are shifted. The behaviors of normal diffusion and CTRW in a regular lattice are similar and the values of MFPT are just shifted in CTRW resulted. In both processes the mutant enzymes outperform EcoRV for small values of the binding constant and for larger values of K_{ns}^0 EcoRV performance is slightly better than the mutant enzyme.

As we had expected, fractional Brownian motion and diffusion on percolation cluster at criticality due to sharing the same nature, have similar behaviors. EcoRV outperforms the mutant enzyme almost for all ranges of the binding constant. We may make this conclusion that CTRW on percolation inherits mostly the fractal nature properties, as it shows the similar behavior to fBm and percolation cluster.

Subdiffusion-limited reactions generally increase the likelihood for biochemical reactions to occur when the reactants are close-by [16, 23]. Such a more local picture of cellular biomolecular reactions in fact ties in with the observed colocalisation of interacting genes [121]. In higher cells the similar locality is effected by internal compartmentalisation by membranes. It will be interesting to obtain more detailed information from single particle tracking experiments in living cells, in order to develop an integrated theory for cellular signalling and regulation under crowding conditions in living cells.

Appendix A

Generating percolation cluster

To give some insight how generating percolation cluster works, two subroutines that generate the percolation cluster and find the largest cluster are displayed.

```
subroutine create_percolation_cluster(idum,pc,numpointx,numpointy,
    numpointz,
bondxplus,bondyplus,bondzplus, bondxminus,bondyminus,bondzminus,nbond)
Implicit None
integer i,j,k,numpointx,numpointy,numpointz
real pc
integer bondxminus(0:numpointx+1,0:numpointy+1,0:numpointz+1)
,bondyminus(0:numpointx+1,0:numpointy+1,0:numpointz+1), &
bondxplus(0:numpointx+1,0:numpointy+1,0:numpointz+1),
bondyplus(0:numpointx+1,0:numpointy+1,0:numpointz+1)
integer bondzminus(0:numpointx+1,0:numpointy+1,0:numpointz+1),
bondzplus(0:numpointx+1,0:numpointy+1,0:numpointz+1)
integer idum
bondxplus=0
bondyplus=0
bondxminus=0
bondyminus=0
bondzplus=0
bondzminus=0
! Array bondxplus (i,j,k) has the value of 1,
if the bond between (i,j,k) to
(i,j+1,k) is available, otherwise it has the valuse of 0.
!The same goes for arrays bondyplus and bondzplus
! Array bondxminus (i,j,k) has the value of 1,
if the bond between(i,j,k) to
!(i,j-1,k) is available, otherwise it has the value of 0.
!The same goes for arrays bondyminus and bondzminus
Do i=1,numpointx
  Do j=1, numpointy
    Do k=1,numpointz
      p=ran3(idum)
        if(p<pc)then
```

```
           bondxplus(i,j,k)=1
         End if
         bondxminus(i,j+1,k)=bondxplus(i,j,k)
       End Do
     End Do
   End Do
   Do j=1,numpointy
     Do i=1, numpointx
       Do k=1,numpointz
         p=ran3(idum)
         if(p<pc) then
           bondyplus(i,j,k)=1
         End if
         bondyminus(i+1,j,k)=bondyplus(i,j,k)
       End Do
     End Do
   END DO
   Do i=1,numpointx
     Do j=1, numpointy
       Do k=1,numpointz
         p=ran3(idum)
         if(p<pc) then
           bondzplus(i,j,k)=1
         End if
         bondzminus(i,j,k+1)=bondzplus(i,j,k)
       End Do
     End Do
   END DO
end subroutine
subroutine  find_infinite_cluster(bondxminus,bondyminus,bondyplus,
   bondxplus,&
bondzminus,bondzplus,numpointx,numpointy,numpointz)
Implicit None  !finding the biggest cluster
integer  ::numpointx,numpointy,numpointz,i,j,ii,jj,k,kk,m,n,o,kn
integer :: numcluster,  nmaxcluster,  sizemax,infiniteclusterindex,newpoint
   ,time
integer ::  bondxminus(0:numpointx+1,0:numpointy+1,0:numpointz+1),
bondyminus(0:numpointx+1,0:numpointy+1,0:numpointz+1), &
bondxplus(0:numpointx+1,0:numpointy+1,0:numpointz+1),&
bondyplus(0:numpointx+1,0:numpointy+1,0:numpointz+1),
bondzminus(0:numpointx+1,0:numpointy+1,0:numpointz+1),
bondzplus(0:numpointx+1,0:numpointy+1,0:numpointz+1)
integer ::clusterindex(0:numpointx+1,0:numpointy+1,0:numpointz+1),
clustersize(0:numpointx*numpointy*numpointz), &
& nbondnew(0:numpointx+1,0:numpointy+1,0:numpointz+1)
integer ::probnew(0:numpointx+1,0:numpointy+1,0:numpointz+1),
probold(0:numpointx+1,0:numpointy+1,0:numpointz+1)
! Array bondxplus (i,j,k) has the value of 1,
if the bond between (i,j,k) to
(i,j+1,k) is available, otherwise it has the valuse of 0.
!The same goes for arrays bondyplus and bondzplus
! Array bondxminus (i,j,k) has the value of 1,
```

```
      if the bond between(i,j,k) to
    !(i,j-1,k) is available, otherwise it has the value of 0.
    !The same goes for arrays bondyminus and bondzminus
    sizemax=0
    clustersize=0
    probold=0
    probnew=0
    clusterindex=0
    numcluster=0
    nbondnew=0
    !finding the biggest cluster
    Do i=1,numpointx
     Do j=1,numpointy
      Do k=1,numpointz
        if(clusterindex(i,j,k)==0)then
          numcluster=numcluster+1
          clusterindex(i,j,k)=numcluster
          do ii=1,numpointx
101         do jj=1,numpointy
            do kk=1,numpointz
              probnew(ii,jj,kk)=0
              !probability for point ii,jj,kk in 3 dimension
              probold(ii,jj,kk)=0
            end do
           end do
          end do
          probold(i,j,k)=1
          newpoint=1
          time=1
          do while(time<=numpointx*numpointy*numpointz.and.newpoint>0)
            newpoint=0
            Do m=1,numpointx
             Do n=1,numpointy
              Do o=1,numpointz
              !calculating the probability from (i,j,k)to go to a new point on
                the lattice
                !that could be (i+1,j,k),(i-1,j,k),(i,j+1,k),(i,j-1,k),(i,j,k
                  +1),(i,j,k-1)
              probnew(m,n,o)=probold(m,n+1,o)*bondxplus(m,n,o)+
              probold(m,n-1,o)*bondxminus(m,n,o)
121           probnew(m,n,o)=probnew(m,n,o)+probold(m+1,n,o)*bondyplus(m,n,o)+
              probold(m-1,n,o)*bondyminus(m,n,o)
              probnew(m,n,o)=probnew(m,n,o)+probold(m,n,o+1)*bondzplus(m,n,o)+
              probold(m,n,o-1)*bondzminus(m,n,o)
                  if(probnew(m,n,o)>1)probnew(m,n,o)=1
                  if(probnew(m,n,o)>0.and.clusterindex(m,n,o)==0)then
                    clusterindex(m,n,o)=numcluster
                    newpoint=newpoint+1
                  end if
               end do
              end do
             end do
```

```
      Do m=1,numpointx
      Do n=1,numpointy
      Do o=1,numpointz
      probold(m,n,o)=probnew(m,n+1,o)*bondxplus(m,n,o)+
      probnew(m,n-1,o)*bondxminus(m,n,o)
      probold(m,n,o)=probold(m,n,o)+probnew(m+1,n,o)*bondyplus(m,n,o)+
      probnew(m-1,n,o)*bondyminus(m,n,o)
      probold(m,n,o)=probold(m,n,o)+probnew(m,n,o+1)*bondzplus(m,n,o)+
141   probnew(m,n,o-1)*bondzminus(m,n,o)
         if(probold(m,n,o)>1)probold(m,n,o)=1
           if(probold(m,n,o)>0.and.clusterindex(m,n,o)==0)then
           clusterindex(m,n,o)=numcluster
           newpoint=newpoint+1
           end if
      end do
      end do
      end do
      time=time+2
      End do
      End if
      End Do
      End Do
      End Do
      nmaxcluster=numcluster
      Do kn=1,nmaxcluster
        do i=1,numpointx
        do j=1,numpointy
        do k=1,numpointz
161     if(clusterindex(i,j,k)==kn)then
        clustersize(kn)=clustersize(kn)+1
        end if
        if(clustersize(kn)>sizemax)then
           sizemax=clustersize(kn)
           infiniteclusterindex=kn
        end if
      end do
      end do
      end do
      End Do
      print*,"sizemax,infiniteclusterindex",sizemax,infiniteclusterindex
      ! deteling smaller clusters, except the biggest cluster
      do i=1,numpointx
        do j=1,numpointy-1
        do k=1,numpointz-1
           if(clusterindex(i,j,k).ne.infiniteclusterindex)then
                bondxplus(i,j,k)=0
                bondxminus(i,j+1,k)=bondxplus(i,j,k)
           end if
181     end do
      end do
      end do
      do i=1,numpointx-1
```

```
      do j=1,numpointy
       do k=1,numpointz-1
        if(clusterindex(i,j,k).ne.infiniteclusterindex)then
             bondyplus(i,j,k)=0
             bondyminus(i+1,j,k)=bondyplus(i,j,k)
        end if
       end do
      end do
     end do
     do i=1,numpointx-1
      do j=1,numpointy-1
       do k=1,numpointz
        if(clusterindex(i,j,k).ne.infiniteclusterindex)then
             bondzplus(i,j,k)=0
             bondzminus(i,j,k+1)=bondzplus(i,j,k)
        end if
       end do
      end do
     end do
     end subroutine
```

Appendix B

Hosking

This subroutine generated fGn using Hosking method. [112]

```
xtern double covariance(long i, double H) {
  if (i == 0) return 1;
  else return (pow(i-1,2*H)-2*pow(i,2*H)+pow(i+1,2*H))/2;
}
void hosking(long *nn, double *H, double *L, int *cum, long *seed1, long
    *seed2, double *output) {
  /* function that generates a fractional Brownian motion or fractional
      */
  /* Gaussian noise sample using the Hosking method.
                                    */
  /* Input:   *n       determines the sample size N by N=2^(*n)
                                    */
  /*          *H       the Hurst parameter of the trace
                                    */
  /*          *L       the sample is generated on [0,L]
                                    */
  /*          *cum     = 0: fractional Gaussian noise is produced
             */
  /*                   = 1: fractional Brownian motion is produced
             */
  /*          *seed1   seed1 for the random generator
                                    */
  /*          *seed2   seed2 for the random generator
                                    */
  /* Output:  *seed1   new seed1 of the random generator
                                    */
  /*          *seed2   new seed2 of the random generator
                                    */
  /*          *output  the resulting sample is stored in this array
                     */
  long i, j, generator, m = pow(2,*nn);
  double *phi = (double *) calloc(m, sizeof(double));
  double *psi = (double *) calloc(m, sizeof(double));
  double *cov = (double *) calloc(m, sizeof(double));
```

```
      double v, scaling;

      /* set random generator and seeds */
      snorm();
      generator = 1;
      gscgn(1, &generator);
      setall(*seed1,*seed2);

      /* initialization */
      output[0] = snorm();
      v = 1;
      phi[0] = 0;
      for (i=0; i<m; i++)
        cov[i] = covariance(i, *H);

      /* simulation */
      for(i=1; i<m; i++) {
        phi[i-1] = cov[i];
        for (j=0; j<i-1; j++) {
          psi[j] = phi[j];
          phi[i-1] -= psi[j]*cov[i-j-1];
        }
        phi[i-1] /= v;
        for (j=0; j<i-1; j++) {
          phi[j] = psi[j] - phi[i-1]*psi[i-j-2];
        }
        v *= (1-phi[i-1]*phi[i-1]);

        output[i] = 0;
        for (j=0; j<i; j++) {
          output[i] += phi[j]*output[i-j-1];
        }
        output[i] += sqrt(v)*snorm();
      }

      /* rescale to obtain a sample of size 2^(*n) on [0,L] */
      scaling = pow(*L/m,*H);
      for(i=0;i<m;i++) {
        output[i] = scaling*(output[i]);
        if (*cum && i>0) {
          output[i] += output[i-1];
        }
      }

      /* store the new random seeds and free memory */
      getsd(seed1,seed2);

      free(phi);
      free(psi);
      free(cov);
    }
```

Appendix C

Waiting time distribution with cutoff

This subroutine generate random numbers taken from distribution given by Eq.(4.25)

```
#define MAXIT 1000
#define UNUSED (-1.11e30)
#define SIGN(a,b) ((b) >= 0.0 ? fabs(a) : -fabs(a))
#define IM1 2147483563
#define IM2 2147483399
#define AM (1.0/IM1)
#define IMM1 (IM1-1)
#define IA1 40014
#define IA2 40692
#define IQ1 53668
#define IQ2 52774
#define IR1 12211
#define IR2 3791
#define NTAB 32
#define NDIV (1+IMM1/NTAB)
#define EPS 1.2e-7
#define RNMX (1.0-EPS)

float zriddr(float (*func)(float,float),float x1,float x2,float rn,
    float xacc);
float dum;
float ctrw(float,float);
float xacc=1e-6; /* Accuracy of Zriddr function*/

float tau=1.0;

float Lmd=0.00001; /* inverse of cutoff time*/

int main(){

  float cfv;

  cfv=1.0-1.0/pow(1.0+RAND_MAX/tau,alpha);
```

```
/*Cumulative function for power law waiting time distribution*/
do{
   dum=ran3(&idum);
   }while (dum>=cfv); /* Generation of a random number smaller than
      cfv */


waitingtime=zriddr(ctrw,0.0,RAND_MAX,dum,xacc);
/*Zriddr finds the root of ctrw function*/
}


float ctrw(float x,float rn)
{
  float ans;
  /*do{
    dum=ran2(&idum);
    }while (dum==1); */

  ans=1.0-exp(-Lmd*x)/pow(1.0+x/t0,alpha)-dum;
  //ans=1.0-1.0/pow(1.0+x/t0,alpha)-rn;
  return ans;
}
float zriddr(float (*func)(float,float),float x1,float x2,float rn,
    float xacc)
{
  int j;
  float ans,fh,fl,fm,fnew,s,xh,xl,xm,xnew;
  fl=(*func)(x1,rn);
  fh=(*func)(x2,rn);
  // printf("x1=%f,x2=%f\n",x1,x2);
  //printf("rn=%f\n",rn);
  //printf(" fl=%f,fh=%f\n",fl,fh);
  if((fl>0.0 && fh<0.0)||(fl<0.0 && fh>0.0)){
    xl=x1;
    xh=x2;
    ans=UNUSED;
    for(j=1;j<=MAXIT;j++){
      xm=0.5*(xl+xh);
      fm=(*func)(xm,rn);
      s=sqrt(fm*fm-fl*fh);
      if (s==0.0) return ans;
      xnew=xm+(xm-x1)*((fl>=fh ? 1.0 : -1.0)*fm/s);
      if (fabs(xnew-ans)<=xacc) return ans;
      ans=xnew;
      fnew=(*func)(ans,rn);
```

```
        if (fnew==0.0) return ans;
        if (SIGN(fm,fnew) != fm){
87  xl=xm;
    fl=fm;
    xh=ans;
    fh=fnew;
        } else if (SIGN(fl,fnew)!=fl){
    xh=ans;
    fh=fnew;
        } else if (SIGN(fh,fnew)!=fh){
    xl=ans;
    fl=fnew;
        } else {
    printf("never get here.");
    exit(1);
        }
        if (fabs(xh-xl)<=xacc) return ans;
      }
      printf("zridder exceed max iterations.");
      printf("random number is %f\n",dum);
      exit(1);
    }
107 else {
      if (fl==0.0) return x1;
      if (fh==0.0) return x2;
      printf("root must be bracketed in zridder.\n");
      printf("random number is %f\n",dum);
      printf("fl=%f and fh=%f\n",fl,fh);
      exit(1);
    }
    return 0.0;
}
```

Bibliography

[1] F. Family and D. P. Landau Kinetics of Aggregation and Gelation, North-Holland, The Netherlands (1984).

[2] S. A. Rice, Diffusion-Limited Reactions, Elsevier, NewYork (1985).

[3] M.V. Smoluchowski, Drei Vorträge über Diffusion, Brownische Molekularbewegung und Koagulation von Kolloidteilchen. Physik. Zeit. **17**: 557, 585 (1916).

[4] P. H. von Hippel and O. G. Berg, Facilitated target location in biological systems, J. Biol. Chem. **264**, 675 (1989).

[5] R. J. Ellis and A. P. Minton, Cell biology: Join the crowd, Nature, London **425**, 27 (2003).

[6] A. P. Minton, How can biochemical reactions within cells differ from those in test tubes?, J. Cell Science **199**, 2863 (2006).

[7] A.P Minton, Excluded volume as a determinant of macromolecular structure and reactivity. Biopolymers **20**, 2093-2120 (1981).

[8] Lebowitz, J. L., Helfand, E. and Praestgaard, E. Scaled particle theory of fluid mixtures. J. Chem. Phys. **43**, 774-779 (1965).

[9] A. P. Minton, The influence of macromolecular crowding and macromolecular confinement on biochemical reactions in physiological media. J. Biol. Chem. **276** (14), 1057780 (2001).

[10] A. P.Minton, Molecular crowding: analysis of effects of high concentrations of inert cosolutes on biochemical equilibria and rates in terms of volume exclusion. Meth. Enzymol. **295**, 127-149 (1998).

[11] S. B. Zimmerman and S. O. Trach, Estimation of macromolecule concentrations and excluded volume effects for the cytoplasm of Escherichia coli, J. Mol. Biol. **222**, 599 (1991).

[12] S. B. Zimmerman and A. P. Minton, Macromolecular crowding: biochemical, biophysical and physiological consequences, Annu. Rev. Biophys. Biomol. Struct. **22**, 27 (1993).

[13] M. S. Cheung, D. Klimov, and D. Thirumalai, Molecular crowding enhances native state stability and refolding rates of globular proteins, Proc. Natl. Acad. Sci. U.S.A. **102**, 4753 (2005).

[14] A. P. Minton, Implications of macromolecular crowding for protein assembly, Curr. Opin. Struct. Biol. **10**, 34 (2000).

[15] S. R. McGuffee and A. H. Elcock, Diffusion, Crowding and Protein Stability in a Dynamic Molecular Model of the Bacterial Cytoplasm, PLoS Comput. Biol. **6**, e1000694 (2010).

[16] I. Golding and E. C. Cox, Physical nature of bacterial cytoplasm, Phys. Rev. Lett. **96**, 098102 (2006).

[17] S. C. Weber, A. J. Spakowitz, and J. A. Theriot, Subdiffusive motion of a polymer composed of subdiffusive monomers, Phys. Rev. Lett. **104**, 238102 (2010).

[18] G. Seisenberger, M. U. Ried, T. Endreß, H. Büning, M. Hallek, and C. Bräuchle, Real-time single-molecule imaging of the infection pathway of an adeno-associated virus, Science **294**, 1929 (2001).

[19] J.-H. Jeon et al., V. Tejedor, S. Burov, E. Barkai, C. Selhuber, K. Berg-Sørensen, L. Oddershede, and R. Metzler, In Vivo Anomalous Diffusion and Weak Ergodicity Breaking of Lipid Granules, Phys. Rev. Lett. **106** 048103 (2011).

[20] M. Weiss, M. Elsner, F. Kartberg, and T. Nilsson, Anomalous subdiffusion is a measure for cytoplasmic crowding in living cells, Biophys. J. **87**, 3518 (2004).

[21] J. Szymanski and M. Weiss, Elucidating the Origin of Anomalous Diffusion in Crowded Fluids, Phys. Rev. Lett. **103**, 038102 (2009).

[22] M. A. Lomholt, I. M. Zaid, and R. Metzler, Subdiffusion and weak ergodicity breaking in the presence of a reactive boundary, Phys. Rev. Lett. **98**, 200603 (2007).

[23] G. Guigas and M. Weiss, Sampling the cell with anomalous diffusion - The discovery of slowness, Biophys. J. **9**, 90 (2008).

[24] H. E. Kubitschek, Cell volume increase in Escherichia coli after shifts to richer media., J. Bacteriol. 172 (1), 94 (1990).

[25] J. Bitinaite, D. A. Wah, A. K. Aggarwal, and I. Schildkraut, FokI dimerization is required for DNA cleavage, Proc Natl Acad Sci USA **95** (18), 10570 (1998).

[26] A. Pingoud and A. Jeltsch, Structure and function of type II restriction endonucleases, Nucleic Acids Research **29** (18), 37053727(2001).

[27] S. G. Erskine, G. S. Baldwin, and S. E. Halford, Rapid reaction analysis of plasmid DNA cleavage by the EcoRV restriction endonuclease, Biochem. **36**, 7567 (1997).

[28] F. K. Winkler, D. W. Banner, C. Oefner, D. Tsernoglou, and R. S. Brown, The crystal structure of EcoRV endonuclease and of its complexes with cognate and non-cognate DNA fragments, EMBO J. **12**, 1781 (1993).

[29] B. v. d. Broek, M. A. Lomholt, S.-M. J. Kalisch, R. Metzler, and G. J. L. Wuite, How DNA coiling enhances target localization by proteins, Proc. Natl. Acad. Sci. USA **105**, 41, 15738 (2008).

[30] D. Ben-Avraham and S. Havlin, Diffusion and Reactions in Fractals and Disordered Systems, Cambridge University Press, Cambridge, UK, (2005).

[31] A. Klemm, R. Metzler, and R. Kimmich, Diffusion on random-site percolation clusters: Theory and NMR microscopy experiments with model objects, Phys. Rev. E **65**, 021112 (2002).

[32] R. Cuthbertson, W. M. L. Holcombe, and R. Paton,Computation in cellular and molecular biological systems,World Scientific, Singapore (1995).

[33] C. D. Lorenz and R. M. Ziff,Precise determination of the bond percolation thresholds and finite-size scaling corrections for the sc, fcc, and bcc lattices, Phys. Rev. E **57**, 230 (1998).

[34] Similar insights into the effect of crowding environments were obtained from lattice random walks and off-lattice Brownian dynamics simulations in J. D. Schmit, E. Kamber, and J. Kondev, Lattice Model of Diffusion Limited Bimolecular Chemical Reactions in Confined Environments, Phys. Rev. Lett. **102**, 218302 (2009);
N. Dorsaz, C. De Michele, F. Piazza, P. de los Rios, and G. Foffi, Diffusion Limited Reactions in Crowded Environments, Phys. Rev. Lett. **105**, 120601 (2010).

[35] C. C. Fritsch and J. Langowski, Anomalous diffusion in the interphase cell nucleus: The effect of spatial correlations of chromatin, J. Chem. Phys. **133**, 025101 (2010).

[36] C. Loverdo, O. Bénichou, and R. Voituriez, Quantifying Hopping and Jumping in Facilitated Diffusion of DNA-Binding Proteins, Phys. Rev. Lett. **102**, 188101 (2009).

[37] M. J. Saxton, Chemically limited reactions on a percolation cluster, J. Chem. Phys. **116**, 203 (2002).

[38] B. Mandelbrot, Self-affine fractals and fractal dimension. Phys. Scr. **32**, 257 260 (1985).

[39] E.W. Montroll and H. Sher, Random walks on lattices. IV. Continuous-time walks and influence of absorbing boundaries, J. Stat. Phys. **9**,101 (1973); H. Scher and E. W. Montroll, Anomalous transit-time dispersion in amorphous solids, Phys. Rev. B **12**,2455 (1975).

[40] A. V. Weigel, Blair Simon, Michael M. Tamkun, and Diego Krap, Ergodic and nonergodic processes coexist in the plasma membrane as observed by single-molecule tracking, Proc Natl Acad Sci U S A **108**, 6438 (2011).

[41] Y. Meroz and I.M. Sokolov, Subdiffusion of mixed origins: when ergodicity and nonergodicity coexist, Phys. Rev. E **81**, 010101 (2010).

[42] http://en.wikipedia.org/wiki/Fractal

[43] La Monadologie, Gottfried Leibniz, (1714).

[44] Clifford A. Pickover, The Math Book: From Pythagoras to the 57th Dimension, 250 Milestones in the History of Mathematics. Sterling Publishing Company, Inc. p. 310 (2009).

[45] J. Perrin, les Atomes Paris 1913, Alcan. A 1970 reprint by Gallimard supersedes several revisions that had aged less successfully. Englsih translation: Atoms by D.L Hammick, London Constable. Newyork van Nostrand Also translated into German Polish Russian Serbian and Japanese (1913).

[46] R. Metzler and J. Klafter, The random walk's guide to anomalous diffusion, a fractional dynamics approach, Physics Reports **339**,1 (2000).

[47] Michael Batty, Fractals - Geometry Between Dimensions. New Scientist (Holborn Publishing Group) **105** 1450, 31 (1985).

[48] John C. Russ, Fractal surfaces, Volume 1. Springer. p. 1.(1994).

[49] B.B.Mandelbrot . The Fractal Geometry of Nature. W.H. Freeman and Company (1982).

[50] Y. Meyer and S. Roques, Progress in wavelet analysis and applications: proceedings of the International Conference Wavelets and Applications,Toulouse, France, p25, June (1992).

[51] S. V. Buldyrev, A. L. Goldberger, S. Havlin, C. K. Peng and H. E. Stanley. chapter 3 in A. Bunde and S. Havlin Eds. Fractals in Science (1995).

[52] J. W. Dollinger, R. Metzler, and T. F. Nonnenmacher, Bi-asymptotic fractals: Fractals between lower and upper bounds, J. Phys. A **31**, 3839 (1998).

[53] Jánosi, Fractal clusters and self-organised criticality, Fractals **2**, 1, 153-168(1994).

[54] A. Bunde, Sh. Havlin, Fractals and Disordered Systems, Springer (1991).

[55] A.Aharony, Fractals in Statistical Physics. Annals of the New York Academy of Sciences, 452, 220-225 (1985).

[56] H. E. Stanley, Cluster shapes at the percolation threshold: and effective cluster dimensionality and its connection with critical-point exponents, J. Phys. A: Math. Gen. 10 L211(1977).

[57] B. D. Hughes, Random Walks and Random Environments, Vol. 1 Oxford University Press, Oxford, UK (1995).

[58] S, Redner, A Guide to First-Passage Processes, Cambridge, Cambridge University Press (2001).

[59] B.V. Gnedenko and A.N. Kolmogorov, Limit Distributions for Sums of Independent Random Variables, Reading, Ma, Addison-Wesley, (1954).

[60] E.W. Montroll, and G.H. Weiss Random walks on lattices-II, Journal of Mathematical Physics, **6**, 167-181 (1965).

[61] C. Monthus and J. P. Bouchaud, Models of traps and glass phenomenology ,J. Phys. A: Math. Gen. **29**, 3847(1996).

[62] E. Barkai and Y.Cheng Aging continuous time random walks, J. Chem. Phys. **118**, 6167 (2003).

[63] J. P. Bouchaud, Weak ergodicity breaking and aging in disordered systems, J. Phys. I France **2**, 9, 1705-1713 (1992).

[64] P. G de Gennes On the relation between percolation theory and the elasticity od gels, J. Phys. Lett., Paris **37**, L1, chapter 6, (1976).

[65] S. Alexander and R. Orbach density of states on fractals, fractons, J. Phys. lett. Paris **43** L625.chap 1, 5, 6 (1982).

[66] Mandelbrot, B., and Wallis, J., Noah, Joseph and operational hydrology, Water Resources Research **4**, 909-918 (1968).

[67] W.Gerstner and W. M. Kistler, Spiking Neuron Models, Cambridge, Cambridge University Press, (2002).

[68] T. Vicsek, Fractal Growth Phenomena, Singapore, World Scientific (1989).

[69] R. Metzler and J. Klafter, When translocation dynamics becomes anomalous, Biophys. J. **85**, 2776 (2003).

[70] G. Pfister and H. Scher, Time-dependent electrical transport in amorphous solids, Phys. Rev. B **15**,2062 (1977).

[71] R. Metzler and J. Klafter, Boundary value problems for fractional diffusion equations, Physica A **278**, 107 (2000).

[72] G. Rangarajan and M. Ding, Fractals **8**, 139 (2000).

[73] M. Ding, W. Yang, Distribution of the first return time in fractional Brownian motion and its application to the study of on-off intermittency, Phys. Rev. E **52**, 207 (1995).

[74] R. Metzler, J Klafter 2000, The random walk's guide to anomalous diffusion: a fractional dynamics approach, Phys. Rep. **339**, 1 (2000).

[75] J.-P. Bouchaud, Anomalous diffusion in disordered media: Statistical mechanisms, models and physical applications, Phys. Rep. **195**, 127 (1990).

[76] I. Nordlund, A new determination of Avogadro's number from Brownian motion of small mercury spherules, Z. f. Phys. Chem **87**, 40 (1914).

[77] A. Caspi, R. Granek, and M. Elbaum, Diffusion and directed motion in cellular transport, Phys. Rev. E **66**, 011916 (2002).

[78] I. Y. Wong, M. L. Gardel, D. R. Reichman, E. R. Weeks, M. T. Valentine, A. R. Bausch, and D. A. Weitz, Anomalous Diffusion Probes Microstructure Dynamics of Entangled F-Actin Networks, Phys. Rev. Lett. **92**, 178101 (2004).

[79] W. Pan, L. Filobelo, N. D. Q. Pham, O. Galkin, V. V. Uzunova, and P. G. Vekilov, Viscoelasticity in Homogeneous Protein Solutions, Phys. Rev. Lett. **102**, 058101 (2009).

[80] S. S. Rogers, C. van der Walle, and T. A. Waigh, Microrheology of Bacterial Biofilms In Vitro: Staphylococcus aureus and Pseudomonas aeruginosa, Langmuir **24**, 13549 (2008).

[81] J.-H. Jeon and R. Metzler, Fractional Brownian motion and motion governed by the fractional Langevin equation in confined geometries, Phys. Rev. E **81**, 021103 (2009).

[82] W. Deng and E. Barkai, Ergodic properties of fractional Brownian-Langevin motion, Phys. Rev. E **79**, 011112 (2009).

[83] A. Lubelski, I. M. Sokolov, and J. Klafter, Nonergodicity Mimics Inhomogeneity in Single Particle Tracking, Phys. Rev. Lett. **100**, 250602 (2008).

[84] Y. He, S. Burov, R. Metzler, and E. Barkai, Random Time-Scale Invariant Diffusion and Transport Coefficients, Phys. Rev. Lett. **101**, 058101 (2008).

[85] Brokmann, X., Hermier, J.-P., Messin, G., Desbiolles, P., Bouchaud, J.-P., Dahan, M.: Phys. Rev. Lett. **90**, 120601 (2003).

[86] G. Margolin, E. Barkai, Nonergodicity of Blinking Nanocrystals and Other Lévy-Walk Processes, Phys. Rev. Lett. **94**, 080601 (2005).

[87] Margolin, G., Barkai, E., Nonergodicity of a Time Series Obeying Lévy Statistics, J. Stat. Phys. **122**, 137 (2006).

[88] G. Bel, E. Barkai, Weak Ergodicity Breaking in the Continuous-Time Random Walk, Phys. Rev. Lett. **94**, 240602 (2005).

[89] E. Barkai, Residence Time Statistics for Normal and Fractional Diffusion in a Force Field, J. Stat. Phys. **123**, 883 (2006).

[90] S. Burov, E. Barkai, Occupation Time Statistics in the Quenched Trap Model, Phys. Rev. Lett. **98**, 250601 (2007).

[91] H. Scher, M. F. Shlesinger, and J. T. Bendler, Time-scale invariance in transport and relaxation, Phys. Today **44**, 1 (1991).

[92] J. Szymanski and M. Weiss, Elucidating the Origin of Anomalous Diffusion in Crowded Fluids, Phys. Rev. Lett. **103**, 038102 (2009).

[93] M. Abramowitz and I.A. Stegun, Handbook of mathematical functions with formulas, graphs, and mathematical tables, Washington (1972).

[94] B. O'Shaughnessy and I. Procaccia, Analytical Solutions for Diffusion on Fractal Objects, Phys. Rev. Lett, **54**, 455 (1985).

[95] Ralf Metzler, Walter G. Glöckle, Theo F. Nonnenmacher, Fractional model equation for anomalous diffusion Physica A **211**, 13 (1994).

[96] Numerical Laplace inversion, Texas A and M University, Harold Vance Department of Petroleum Engineering (2002).

[97] S. Condamin, V. Tejedor, R. Voituriez, O. Bénichou, and J. Klafter, Probing microscopic origins of confined subdiffusion by first-passage observables, Proc. Nat. Acad. Sci. **105**, 5675 (2008).

[98] Kolesov, G., Wunderlich, Z., Laikova, O. N., Gelfand, M. S., and Mirny, L. A., How gene order is in by the biophysics of transcription regulation. PNAS, **104**, 13948 13953 (2007).

[99] J. Hoshen and R. Kopelman, Percolation and cluster distribution. I. Cluster multiple labeling technique and critical concentration algorithm, Phys. Rev. B **14**, 3438 (1976).

[100] I. Majid, D. Ben-Avraham, S. Havlin. H.E. Stanley, Exact-enumeration approach to random walks on percolation clusters in two dimensions, Phys. Rev. B **30**, 1626 (1984).

[101] A. D'Arcy et al., R. S. Brown, M. Zabeau, R. W. v. Resandt, and F. K. Winkler, Purification and crystallization of the EcoRV restriction endonuclease, J. Biol. Chem. 260, 1987 (1985).

[102] http://www.shardcore.org/shardpress/index.php/2006/01/19/e-coli-2006/

[103] http://en.wikipedia.org/wiki/EcoRV

[104] I. Bonnet et al., Sliding and jumping of single EcoRV restriction enzymes on non-cognate DNA, Nucleic Acids Res. **36**, 4118 (2008).

[105] J. Elf, G.-W. Li, X. S. Xie, Probing transcription factor dynamics at the single-molecule level in a living cell, Science **316**, 1191 (2007).

[106] A. Jeltsch and A. Pingoud, Kinetic characterization of linear diffusion of the restriction endonuclease Eco RV on DNA, Biochemistry **37**, 2160 (1998).

[107] J.-H. Jeon, A. V. Chechkin, and R. Metzler, First passage behaviour of fractional Brownian motion in two-dimensional wedge domains, Europhys. Lett. **94**, 20008 (2011).

[108] J.B. Bassingthwaighte and Gary M. Raymond, Evaluation of the dispersional analysis method for fractal time series, Annals of Biomedical, Engineering **23**, 491 (1995).

[109] P. Biswas and B.J. Cherayil, Dynamics of fractional brownian walks, J phys chem **99**, 816 (1995).

[110] G. Guigas, C. Kalla, and M. Weiss, Probing the nano-scale viscoelasticity of intracellular fluids in living cells. Biophys. J. **93**, 316 (2007)

[111] J. R. M. Hosking, Modeling persistence in hydrological time series using fractional differencing, Water Resource Res. **20**, 1898 (1984)

[112] http://www2.isye.gatech.edu/ adieker3/fbm/hosking.c

[113] M. V. Berry and Z. V. Lewis, On the Weierstrass-Mandelbrot fractal function. Proc. R. Soc. (Lond.) A. **370**, 459484 (1980).

[114] M.J. Saxton, Anomalous Subdiffusion in Fluorescence Photobleaching Recovery: A Monte Carlo Study, Biophys. Journal **81**, 2226 (2001).

[115] H. Qian, Single-particle tracking: Brownian dynamics of viscoelastic materials. Biophys. J. **79**, 137143 (2000).

[116] T. Dieker, Simulation of fractional Brownian motion, Report, University of Twente, Netherlands (2004)

[117] I. M. Sokolov (person-al communication).

[118] M. Magdziarz, A. Weron,K. Burnecki,J. Klafter. Fractional Brownian motion versus the continuous-time random walk: A simple test for subdiffusive dynamics. Phys Rev Lett. **103**, 180602 (2009).

[119] M. Magdziarz,J. Klafter. Detecting origins of subdiffusion: p-variation test for confined systems. Phys Rev E. **82**, 011129 (2010).

[120] For a recent summary, see, e.g., S. Burov, J.-H. Jeon, R. Metzler, and E. Barkai, Phys. Chem. Chem. Phys. **13**, 1800 (2011).

[121] G. Kolesov et al., Z. Wunderlich, O. N. Laikova, M. S. Gelfand, and L. A. Mirny, How gene order is influenced by the biophysics of transcription regulation, Proc. Natl. Acad. Sci. USA **104**, 13948 (2007).

Publications

[1] M. Neek-Amal, G. Tayebirad, M. Molayem, M. E. Foulaadvand, L. Esmaeili Sereshki, A. Namiranian: "Ground state study of simple atoms within a nano-scale box", Solid State Communications, **145**, 594 (2008).

[2] L.E. Sereshki, M. A. Lomholt and R. Metzler, "A solution to the subdiffusion-efficiency paradox: Inactive states enhance reaction efficiency at subdiffusion conditions in living cells ", EPL **97** 20008(2012).

[3] L.E. Sereshki, M. A. Lomholt and R. Metzler ,"Modeling EcoRV subdiffusion in E.coli", in preparation.

i want morebooks!

Buy your books fast and straightforward online - at one of world's fastest growing online book stores! Environmentally sound due to Print-on-Demand technologies.

Buy your books online at
www.get-morebooks.com

Kaufen Sie Ihre Bücher schnell und unkompliziert online – auf einer der am schnellsten wachsenden Buchhandelsplattformen weltweit! Dank Print-On-Demand umwelt- und ressourcenschonend produziert.

Bücher schneller online kaufen
www.morebooks.de

VDM Verlagsservicegesellschaft mbH
Heinrich-Böcking-Str. 6-8
D - 66121 Saarbrücken

Telefon: +49 681 3720 174
Telefax: +49 681 3720 1749

info@vdm-vsg.de
www.vdm-vsg.de

Printed by Books on Demand GmbH, Norderstedt / Germany